The U.S. City in Transition

Barbara Hahn

The U.S. City in Transition

Springer

Barbara Hahn
Institut für Geographie und Geologie
Universität Würzburg
Würzburg, Germany

This book is a translation of the original German edition „Die US-amerikanische Stadt im Wandel" by Hahn, Barbara, published by Springer-Verlag GmbH, DE in 2014. The translation was done with the help of artificial intelligence (machine translation by the service DeepL.com). A subsequent human revision was done primarily in terms of content, so that the book will read stylistically differently from a conventional translation. Springer Nature works continuously to further the development of tools for the production of books and on the related technologies to support the authors.

ISBN 978-3-662-64863-6   ISBN 978-3-662-64861-2   (eBook)
https://doi.org/10.1007/978-3-662-64861-2

This Springer imprint is published by the registered company Springer-Verlag GmbH, DE, part of Springer Nature.
The registered company address is: Heidelberger Platz 3, 14197 Berlin, Germany

Contents

5.4.3   Evaluation 122
5.5   **Detroit: A Doomed City?** 123
5.5.1   City of the Automobile 123
5.5.2   Population and Building Stock.. 124
5.5.3   Revitalisation Efforts 126
5.5.4   Outlook 127
5.6   **Miami: Economic Recovery of a Polarised City** 128
5.6.1   Population Structure... 128
5.6.2   Post-industrial *Global City* 130
5.6.3   Tourism 131
5.6.4   Future Prospects... 132
5.7   **Atlanta: The Suburban Space as a Location for Services** 132
5.7.1   Upswing to a Service Center.. 132
5.7.2   Decentralisation.. 133
5.7.3   City Center 136
5.8   **New Orleans After Hurricane Katrina** 136
5.8.1   Hurricane Katrina 137
5.8.2   Reconstruction.. 139
5.8.3   Economy... 141
5.8.4   Evaluation 141
5.9   **Phoenix: A Paradise for Seniors?** 142
5.9.1   Sun City 142
5.9.2   Sun City as a Role Model 145
5.9.3   Criticism 145
5.10   **Seattle: High-Tech Location on the Pacific Ocean** 146
5.10.1   Rise to a High-Tech Region 146
5.10.2   Diversification.. 148
5.10.3   Outlook 149
5.11   **Las Vegas Between Hyperreality and Bitter Reality**... 150
5.11.1   Hyperreal Worlds 152
5.11.2   Misery 153

6   **The Future of the American City** 157

         Supplementary Information
         References.. 162
         Index 171

List of Abbreviations

Abbreviations

BID	*Business Improvement District*
BRT	Bus Rapid Transit Line
CBD	*Central Business District*
CDGB	*Community Development Block Grant*
CID	*Common Interest Development*
CMSA	*Consolidated Metropolitan Statistical Area*
CSA	*Combined Statistical Area*
FAR	*Floor Area Ratio*
FEMA	Federal Emergency Management Agency
FHA	Federal Housing Administration
FHWA	Federal Highway Administration
FIRE	*Finance, Insurance, Real Estate*
FTAA	Free Trade Area of the Americas
HOA	Federal Housing Administration
MA	Metropolitan Area
M.I.T.	Massachusetts Institute of Technology
MSA	*Metropolitan Statistical Area*
NAACP	National Association for the Advancement of Colored People
NGO	Nongovernmental Organization
NRHP	National Register of Historic Places
NYCHA	New York City Housing Authority
PMSA	*Primary Metropolitan Statistical Area*
SCHOA	Sun City Home Owners Association
UAW	United Auto Workers
UC	*Urban Cluster*
UGB	*Urban Growth Boundary*
VTA	Valley Transit Authority

US States

AK	Alaska
AL	Alabama
AR	Arkansas
AZ	Arizona
CA	California
CO	Colorado
CT	Connecticut
DE	Delaware
FL	Florida
GA	Georgia
HI	Hawaii
IA	Iowa
ID	Idaho
IL	Illinois
IN	Indiana
KS	Kansas
KY	Kentucky
LA	Louisiana
MA	Massachusetts
MD	Maryland
ME	Maine
MI	Michigan
MN	Minnesota
MO	Missouri
MS	Mississippi
MT	Montana
NC	North Carolina
ND	North Dakota
NE	Nebraska
NH	New Hampshire
NJ	New Jersey

NM	New Mexico	TN	Tennessee
NV	Nevada	TX	Texas
NY	Big Apple		
		UT	Utah
OH	Ohio		
OK	Oklahoma	VA	Virginia
OR	Oregon	VT	Vermont
PA	Pennsylvania	WA	Washington
		WI	Wisconsin
RI	Rhode Island	WV	West Virginia
		WY	Wyoming
SC	South Carolina		
SD	South Dakota		

List of Figures

The U.S.-American City as an Object of Research

Contents

© The Author(s), under exclusive license to Springer-Verlag GmbH, DE, part of Springer Nature 2022
B. Hahn, *The U.S. City in Transition*, https://doi.org/10.1007/978-3-662-64861-2_1

1

Cities are subject to constant change. European cities, quite a few of which were founded as early as the Middle Ages, have experienced several phases of boom and bust in their history. At the same time, the urban system, which encompasses all relationships between cities such as material and immaterial flows, changed. In the United States, cities did not emerge until the eighteenth century on the East Coast and, in some cases, much later in the rest of the country. In many cases, by the mid-twentieth century, the boom was replaced by an abrupt downturn, the end of which is still uncertain. At the same time, other cities were just entering the growth phase, which is still ongoing. Today, prosperous cities are contrasted by cities with completely disastrous development.

With industrialization, many U.S. cities experienced tremendous population growth in the nineteenth century; some numbered more than one million residents just a few decades after their founding. Most cities that owed their boom to textile or heavy industry have experienced sharp population declines in recent decades as a result of deindustrialization and suburbanization, while others have done comparatively well with structural change. In the south and west of the country, many cities were only founded in the course of the twentieth century and grew at a similar rate to the industrial cities in the nineteenth century. The engine of development was now the burgeoning service sector or high-tech industry, which resulted in a fundamental change in the urban system in just half a century. Since the newer cities were founded when the private car had already become a means of mass transport, they differ from the older cities in many respects. Regardless of when they were founded, however, all cities in recent decades have been shaped by suburbanization, decentralization, fragmentation, globalization, privatization, *gentrification,* and decay in at least some areas. These processes have been associated with a change in function.

An anti-urban attitude has always been strongly prevalent in the U.S. But in view of the serious problems of cities, which became increasingly obvious in the mid-twentieth century, and the skyrocketing population losses, in 1962 the long-time head of New York City planning Robert Moses published an essay entitled *Are Cities Dead?.* As Moses had been responsible for the structural development of New York for several decades, he answered the question of the end of the city in the negative, as might be expected, but this opinion did not really enjoy consensus. A commission appointed by President Jimmy Carter (1977–1981) even concluded that urban decline was unstoppable (Lees et al. 2008, p. xvii). The emptying and loss of importance of cities was even encouraged by policy. Cheap gasoline, subsidies for highway construction and maintenance, tax breaks for homeowners, and better schools encouraged suburban

living at the expense of taxpayers and urban renters. As the signs of globalization became more apparent and changes in communications slowly became apparent, the idea of dissolving cities took on new life. Why live in a densely built-up and ecologically polluted city when you can also live in scenic surroundings?

Today, there is no doubt that the U.S. city is by no means dead. Even cities that still showed great signs of decay in the 1980s and hardly gave any reason to believe in a positive development are hardly recognizable. Neighborhoods that one would have given a wide berth to not so long ago have become sought-after addresses. Renovated facades can be seen behind well-kept front gardens, new residential buildings have been erected in vacant lots, and a Starbucks invites you to linger at the street intersections. Where recently there were unkempt high-rise apartment buildings erected in the 1950s–1970s as part of the social housing program, there are parks, shopping centers, or perhaps just unsightly brownfields. In the inner cities, tourists and locals alike stroll along redeveloped lakeshores and riverbanks long dominated by industrial facilities, and new high-rise office and residential buildings, museums, sports stadiums, and shopping malls are being built. New high-income and consumer-oriented urbanites are visibly taking over the cities and use them differently than industrial workers did. The downward trend seems to be broken: Things are looking up again for the U.S. city. But that is only half the truth. While many cities are experiencing a revitalization that would have been unthinkable in the 1990s, other cities are still going downhill. The population continues to decline, dilapidated industrial plants, condemned apartment buildings, empty office buildings, and unkempt brownfields offer little hope of an imminent recovery. The white middle and upper classes have long since abandoned these cities. The former residential areas of the workers are characterized by poverty and, unfortunately, often by crime and drug trafficking. Since tax revenues are low, the room for manoeuvre is limited.

All in all, however, a paradigm shift is emerging, as the value of cities is increasingly being recognized, or at least discussed. In September 2011, the magazine Scientific American presented the future of the city as thoroughly rosy in a themed issue. Cities are portrayed as environmentally friendly, innovative and efficient. City dwellers are seen as more creative and responsible for the country's economic growth to a far greater degree than suburban or rural residents (Glaeser 2011b). A city dweller's environmental footprint is comparatively small, as they are less likely to drive their own car and more likely to walk or use public transport. Cities are extremely productive. If the population doubles, wages and the number of patents per capita increase by 15%. It does not matter whether the population increases from

40,000 to 80,000 or from four to eight million. Another advantage is that when the population grows, the infrastructure, such as the length of roads and supply lines or the number of gas stations, does not increase at the same rate. The more people live in a city, the more efficiently the infrastructure is used (Bettencourt and West 2011, p. 52).

While on the one hand cities are enjoying renewed popularity, suburban sprawl continues unabated. A European flying over the Los Angeles, Houston, or Atlanta metropolitan area for the first time is astonished to see a never-ending patchwork of disparate uses, seemingly haphazardly juxtaposed but always separated by wide streets. Cities are becoming more and more sprawling. This is also the feeling one gets when exploring the cityscapes by car. Now you can also see that the suburban space is by no means as homogeneous or boring as is often claimed, because new concepts have sometimes been implemented in the construction of new housing estates, industrial plants, offices or shopping centers. But the U.S. city is always a fragmented and segmented city. Extremely prosperous parts of the city are juxtaposed with completely run-down neighborhoods, for whose future there is little hope. But who knows?

For the first time in half a century, Barack Obama who lived in Chicago, a native of a big city became president of the United States in 2009. The Obama administration announced during the inauguration that expanding urban infrastructure and improving educational facilities would be important goals in the coming years. However, past experience has repeatedly shown that while programs and money from Washington can help revitalize cities, success ultimately depends on the players on the ground. Unfortunately, not as much money as initially planned could be made available for urban redevelopment either, as the restructuring of the floundering American economy under President Obama took precedence. U.S. cities are all the same only from a distance, in fact they differ from each other in many ways. What is certain is that they have all changed in recent decades, albeit very differently. Moreover, core cities and suburban areas have converged. A clear hierarchical order, which was expressed in the terms urban, suburban, exurban and rural, no longer exists (Lang 2003, p. 29). Geographers, planners, politicians and journalists have identified different trends on which there is no consensus and which cannot be described in conventional terms (❑ Fig. 1.1). Each group as well as many individuals have developed their own vocabulary to illustrate the new processes or to associate them with their own names for all eternity. It is futile to define and delimit the individual terms precisely, as they are used very differently by individual authors. In the following, therefore, only those terms will be used on whose defi-

nition there is a broad consensus, which have become established in general linguistic usage or which have been repeatedly discussed in the scientific literature.

This book focuses on the question of the most important processes and the divergent development of individual cities or parts of cities: Why were some American cities or individual parts of cities more affected by deindustrialization than others? Why have some cities experienced positive development in recent years, while elsewhere the downward spiral that began in the 1950s seems unstoppable? Why, after years of decline, are certain cities or neighborhoods once again attractive to residents and visitors, while others continue to be shunned? Has the upward trend been accompanied by a change in function, or have cities been able to revive their former functions? Why has the population of some cities in the south and west of the country multiplied in a few decades, while other cities in these regions have grown far more slowly? How have the cities changed and how should the changes be assessed? Which groups of the population benefit from positive changes? Are there also losers, and who are they? These and many other fascinating questions are to be answered in what follows. Although the focus of the book is clearly on the core cities and especially on their central areas, the suburban space, without which the U.S. city is inconceivable, must not be neglected.

1.1 The City in Figures

In the U.S., there are no population registration offices that collect data on the population living in a city or region. Censuses, which have been conducted every 10 years since 1790, compensate for this shortcoming. The census is far less controversial in the USA than in Germany and is always accompanied by a great deal of publicity and extensive press coverage (❑ Fig. 1.2). The cut-off date for the last census was April 1, 2010. Every 5 years, i.e. in years ending with a two or seven, the U.S. Bureau of the Census also collects data on the administrative structures of all states. A look at these statistics shows that the U.S. is not only divided into 50 states, but also into 3034 counties, 35,886 municipalities of various legal forms and another 50,087 administrative districts for special tasks. The latter include, among others, 12,884 school districts as well as districts for the water supply of the country (data for 2012, ▶ www.census.gov). The boundaries of special purpose administrative districts do not coincide with municipal boundaries. A school district may include several municipalities as well as land that is not municipally owned. At the same time, larger cities may be assigned to multiple school districts. The high degree of fragmen-

1

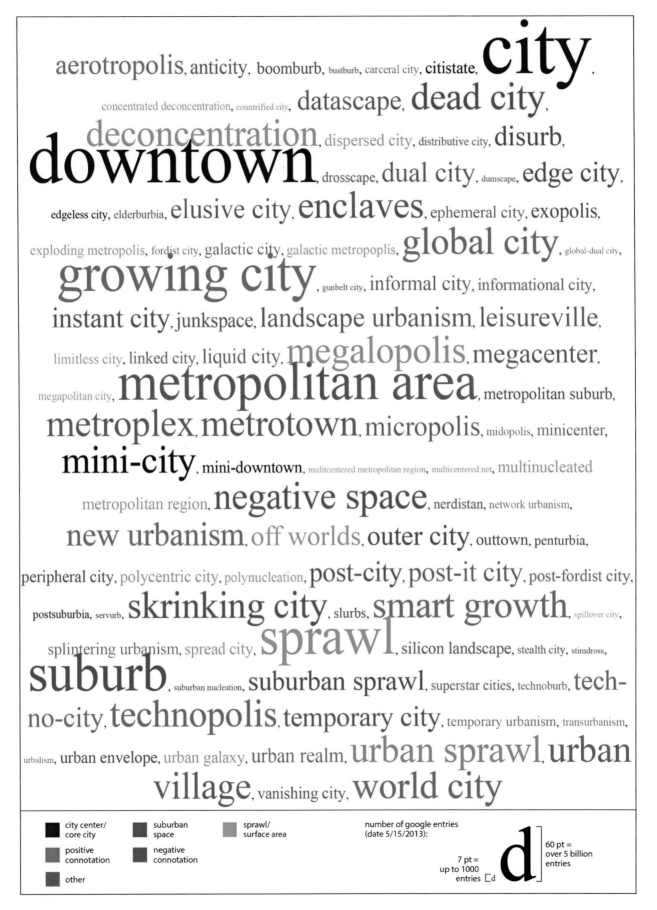

aerotropolis, anticity, boomburb, bustburb, carceral city, citistate, city, concentrated deconcentration, countrified city, datascape, dead city, deconcentration, dispersed city, distributive city, disurb, downtown, drosscape, dual city, dumscape, edge city, edgeless city, elderburbia, elusive city, enclaves, ephemeral city, exopolis, exploding metropolis, fordist city, galactic city, galactic metropoplis, global city, global-dual city, growing city, gunbelt city, informal city, informational city, instant city, junkspace, landscape urbanism, leisureville, limitless city, linked city, liquid city, megalopolis, megacenter, megapolitan city, metropolitan area, metropolitan suburb, metroplex, metrotown, micropolis, midopolis, minicenter, mini-city, mini-downtown, mulitcentered metropolitan region, multicentered net, multinucleated metropolitan region, negative space, nerdistan, network urbanism, new urbanism, off worlds, outer city, outtown, penturbia, peripheral city, polycentric city, polynucleation, post-city, post-it city, post-fordist city, postsuburbia, servurb, skrinking city, slurbs, smart growth, spillover city, splintering urbanism, spread city, sprawl, silicon landscape, stealth city, stimdross, suburb, suburban nucleation, suburban sprawl, superstar cities, technoburb, tech-no-city, technopolis, temporary city, temporary urbanism, transurbanism, urbalism, urban envelope, urban galaxy, urban realm, urban sprawl, urban village, vanishing city, world city

■ city center/core city	■ suburban space	■ sprawl/surface area
■ positive connotation	■ negative connotation	
■ other		

number of google entries (date 5/15/2013):

7 pt = up to 1000 entries ⌈d

d ⌉ 60 pt = over 5 billion entries

■ **Fig. 1.1** Terms relating to the city and suburban space

☐ **Fig. 1.2** Advertising for the 2010 census at the Caribbean Carnival in the New York borough of Brooklyn

tation of public functions is a major reason why large metropolitan areas (see below) have limited governance. Moreover, it is not always easy to allocate the extensive statistical material correctly. On the positive side, however, all data collected by the U.S. Census Bureau are considered public property and can be freely accessed on the Internet (► www.census.gov).

In the USA, municipalities are incorporated and thus become cities when they reach a certain number and density of inhabitants. The values vary from state to state and have been changed several times over time. New cities are given the right to elect their own mayor and other officers, and to set local levies such as sales taxes and property taxes. At the same time, they must assume certain responsibilities, such as funding, managing, and overseeing local schools, garbage collection, fire, and police, that were previously performed by counties. Cities with declining populations can have their title revoked if their population falls below a certain level and density and they are no longer able to perform their municipal functions.

The U.S. Bureau of the Census has repeatedly modified the definition of *urban*. Until 1950, it classified the population of all incorporated communities with more than 2500 residents as urban. Based on prior rights, the population of some smaller New England communities was also included. As population densities outside of cities increased with continued suburbanization, the term *"urbanized area"* was introduced in 1950. In subsequent decades, the use of geographic information systems allowed for a more refined calculation of population density. While the concept of urbanized areas was retained, these are now composed of densely populated regions with populations of 50,000 or more and urban clusters (UCs) with populations of 2500 or more that are also densely populated (U.S. Census Bureau 2011a,

b, p. 3). Although the data cannot be compared exactly due to modified collection methods, there is no doubt that the proportion of the U.S. population living in urban areas has steadily increased. In 1920, for the first time, slightly more than half of all Americans lived in urbanized areas; by 1990, the figure was about three-quarters, and by 2010, it was 80.7%. From 1950 to 2010, the number of cities with at least one million residents increased from five to nine. As Chicago and Philadelphia lost many residents while the American population more than doubled, the percentage of those living in cities with millions increased only from 11.5% to 12.3%. The number of cities with more than 100,000 inhabitants increased significantly from 106 to 273. Whereas in 1950 only 29.4% of all Americans lived in large cities, by 2010 this figure had risen to 43.7% (U.S. Census Bureau: Statistical Abstract 1960, 1991, 2011b; U.S. Census Bureau 2012).

Census tracts are the smallest spatial unit for which data are collected. In 2010, there were 72,531 census tracts with an average of 4256 residents, although there was wide variation: Camp Pendleton Naval Base north of San Diego, which is a separate census tract, had more than 37,000 people; but 90% of census tracts have populations between 1500 and 7500 (Glaeser and Vigdor 2012, p. 5). Most census tracts are socioeconomically and structurally homogeneous neighborhoods and form individual neighborhoods. However, residential-only neighborhoods and adjacent thoroughfares with a divergent population structure are often mapped together in a census tract. U.S. cities are composed of a large number of very different neighborhoods. Neighborhoods are largely homogeneous with a dominant ethnicity and similar income and lifestyle of residents. Neighborhoods can be distinguished from neighboring neighborhoods with a different socioeconomic structure at first glance by different lot sizes or building stock, a different state of maintenance of the houses, and divergent land use. In addition, there are neighborhoods in every city that have experienced negative development in recent years, while others have been able to develop extremely positively.

1.1.1 Population Development Since 1950

The year 1950 represents a turning point in U.S. urban history. Many cities in the old industrialized northeast of the country reached their population peak at that time, while the cities in the south and west recorded large population gains in the following decades. Due to the positive population growth and economic boom of these regions, they are collectively referred to as the sunbelt, in reference to the mostly good weather, and the old-industrialized Northeast, which has very low temperatures in winter, is referred to as the frostbelt. Air-conditioning, invented in an initially primitive form in the 1930s, played an important role in making life and work more comfortable, and thus in boosting the population of cities in the hot south and dry west of the country. The hot and humid climate of Miami or the high temperatures of the desert cities of Las Vegas or Phoenix did not become bearable until the 1950s (Konig 1982, p. 22). New York City was able to maintain its position as the largest city in the United States, even reaching an all-time high of 8.175 million residents in 2010. However, the city was hardly larger at that time than it was in the mid-twentieth century due to population losses from the 1950s through the 1980s. Also, the increase of 167,000 residents in the last decade was rather disappointing, as an increase of more than 400,000 had been expected before the April 1, 2010 census. Of the ten largest cities in 1950, only Los Angeles, located in the western United States, saw a significant increase in population from just under two million to about 3.8 million. By the 1980s, Los Angeles had overtaken Chicago to become the second largest city in the country. More recently, however, Los Angeles' population growth has slowed. In 2010, the city did have 97,000 more residents than 10 years earlier; but in every preceding decade since 1910, Los Angeles had gained more residents. The cities of Chicago, Philadelphia, Detroit, Baltimore, Cleveland, St. Louis, Washington, D.C., and Boston peaked in population in 1950; Detroit, Cleveland, and St. Louis actually lost well over half their populations between 1950 and 2010. The losses occurred because of migration to suburban and sunbelt areas and were hardly interrupted by periods of recovery. Chicago's population had increased by 113,000 during the 1990s. But hopes for a sustained increase in the population of what was once the nation's second-largest city were dashed with the 2010 census, which certified a decline of more than 200,000 residents. Washington, D.C., and Philadelphia enjoyed modest population increases in the first decade of the new millennium after decades of large losses; whether that trend will last remains to be seen. In 2010, only four of the most populous cities in 1950 – New York, Los Angeles, Chicago and Philadelphia – were still among the ten largest cities in the USA. Houston, located in Texas, overtook Philadelphia in the 1980s to become the fourth largest city. The other newcomers Phoenix, San Antonio, San Diego, Dallas and San Jose on the list of the ten largest cities are also located in the sunbelt. Phoenix and San Jose even managed to increase their population more than tenfold in just 60 years (◘ Fig. 1.3).

Similarly, many of the smaller cities in the northeastern United States have lost population in recent decades, while most cities in the south and west of the country have experienced strong population growth. Overall, the development of cities has varied greatly in recent decades. A large number of shrinking cities in the Northeast contrast with many very fast-growing cities in the South and West of the country. There are exceptions, however, in the northeast in the environs of the major cities, where many communities have grown a great deal due to suburbanization. In addition, some sunbelt cities that experienced early industrialization have lost residents. This is true, for example, of the former steel hub of Birmingham, Alabama, whose population has shrunk by about one-third since 1950 to 212,000 in 2010 (U.S. Census Bureau 2011b, 2012).

More recently, the trends of previous decades have become more entrenched. In the 1990 census, metropolitan areas (MAs) replaced the earlier distinction into MSAs (metropolitan statistical areas), CMSAs (consolidated metropolitan statistical areas) and PMSAs (primary metropolitan statistical areas) for the first time and contributed to a simplification in the classification of spatial planning categories. The delimitation of MAs is based on around 2000 different indicators and can only be understood by experts. In simple terms, MAs are made up of core cities and the functionally closely interlinked surrounding area, known as the suburban area. MAs can have one or more core cities and largely correspond to the densification areas of German spatial planning. All regions outside MAs are referred to as rural (ländlich) (U.S. Census Bureau 2011a, pp. 3–4). In 2010, there were 51 MAs in the U.S. with more than one million inhabitants. The ten MAs with the highest population growth from 2000 to 2010 were all in the sunbelt. The Las Vegas, Raleigh, Austin, Charlotte, Riverside-San Bernardino, Orlando, Phoenix, Houston, San Antonio, and Atlanta MAs had grown by at least 20%, twice as much as the 51 MAs on average (Kotkin and Cox 2011).

The 51 MAs with more than one million inhabitants in 2010 have grown in particular due to large population gains in suburban areas. In the 1990s, core cities had contributed 15.4% to the growth of MAs. Since this was the highest in decades, many policymakers and scholars had hoped for a renaissance of the core city and a slow decline in the importance of the suburban area. Those hopes did not materialize, as from 2000 to 2010, the suburban area accounted for 91.4% of MA growth and core cities only 8.6%. Only three core cities gained more residents than the suburban area. In Boston and Providence, this was due to only modest suburban population growth, and in Oklahoma City, suburban-style housing developments had sprung up on large tracts of previously undeveloped land (Kotkin and Cox 2011). The growth of many cities and MAs is not based on a high birth rate of the resident population, but is due to in-migration predominantly from other cities or MAs and from abroad. Rural-urban migration is hardly significant. Analogously, out-migration is responsible for

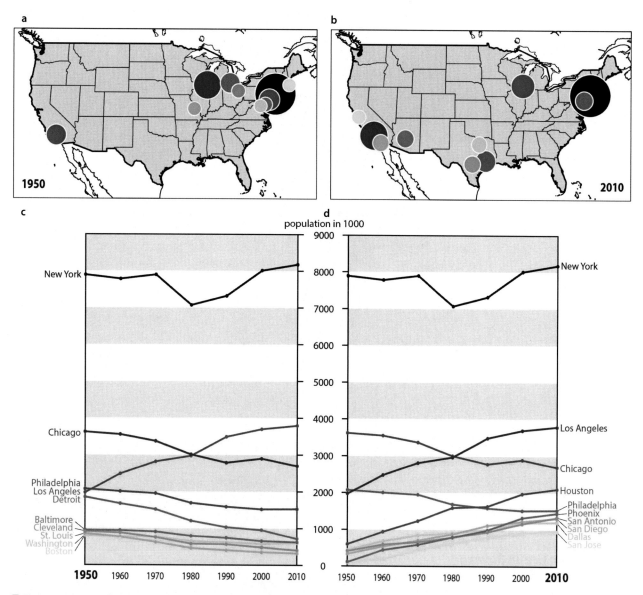

◘ Fig. 1.3 **a** Population development of the ten largest cities in 1950; **b** The ten largest cities in the USA in 1950; **c** Population development of the ten largest cities in 2010; **d** The ten largest cities in the USA in 2010. (Data basis ► www.census.gov)

the declining population figures in cities (Brake and Herfert 2012, p. 13).

In most MAs, population gains have been greatest in the outermost suburban ring for decades. This trend is still continuing. Chicago's inner *suburbs* grew by only 1%, or 30,000 residents, from 2000 to 2010, while the outer *suburbs have seen* new residents increase by more than 500,000, or 16%, since 2000. Only 28% of Chicago MA's nearly 9.5 million residents now live in the core city. The Chicago metropolitan area now occupies a larger area than the Los Angeles MA and ranks third in the world after New York and Tokyo (Cox 2011b). New York City and its environs make up the most populous metropolitan area in the U.S., with nearly 19 million people. Within the core city, Staten Island (Richmond

County), which is characterized by a suburban character, and the Bronx, with 3.9%, experienced the largest gains from 2000 to 2010. Manhattan's population increased by 49,000 to 1,586,000. However, this figure was only about two-thirds of its peak of 2.32 million in 1910. Brooklyn (Kings County) remains New York's most populous borough, with 2.5 million residents, but only managed to grow by 1.6%. As the suburban area of MA New York has grown much faster than the core city in the first decade of the new millennium, its share of the city region's population has declined. New York MA includes 23 counties and grew by 574,000 residents in 10 years (◘ Fig. 1.4). The suburbs accounted for 71% of the region's growth. This was far more than in the 1990s and 1980s, when they accounted for 54% and 48%

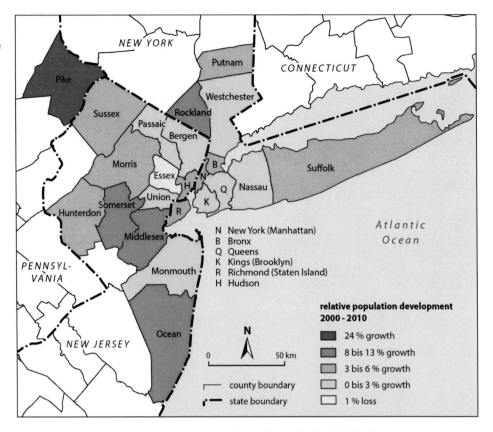

Fig. 1.4 New York: Relative population change 2000–2010. (Data base Cox 2011b, map base ▶ www.census.gov)

of MA's population growth, respectively. From 2000 to 2010, the outer suburbs grew by 6.3%, but the inner suburbs grew by only 2.4%. Thus, there is an increasing redistribution of population to the surrounding areas (Cox 2011c).

1.1.2 Ethnic Breakdown

Immigrants to the United States preferred to settle in large cities until well into the twentieth century because they offered the best opportunities for employment. Often, many immigrants from a particular country settled in one city at times. In Boston, during the potato crisis of the mid-nineteenth century, many Irish arrived by ship, too poor to move on to other locations, while the slaughterhouses of Chicago attracted many Eastern Europeans. The descendants of the nineteenth century immigrants have held American passports for decades or have moved on to other regions of the country. Other cities were preferred destinations for the immigration of blacks from the southern United States. This was especially true in Detroit, where the automobile industry acted as a magnet. In 2010, 12.4% of the U.S. population was foreign-born. In the gateway cities of New York and Los Angeles, but also in San Jose, this was true for well over a third of the population (▶ Fig. 1.5). In San Diego, about a quarter were foreign-born and in Chicago, about

Fig. 1.5 Brighton Beach in New York is a traditional location for Russian immigrants

a fifth. The rate was particularly low, understandably, at only 6.5% in Detroit, which is old industrialized and has been shrinking rapidly for decades; it is surprising, however, that the figure in the boomtown of Atlanta is barely higher, at 7.8%. In recent decades, hispanics in particular have immigrated to the U.S., i.e. people from Mexico, the Caribbean and Central and South America. In 2010, 16.3% of all people living in the U.S. identified themselves as Hispanic. Their percentage is far higher

■ **Fig. 1.6** Ethnic composition of the population of selected cities in 2010. (Data basis ► www.census.gov)

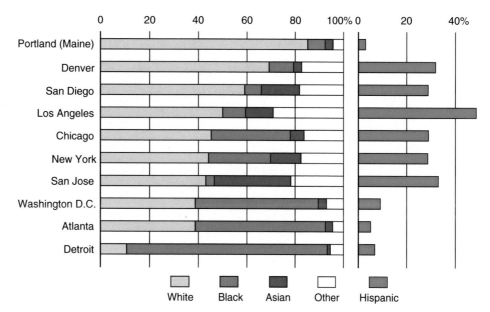

White Black Asian Other Hispanic

in most cities than in suburban or rural areas due to the large supply of unskilled workers. Interestingly, in San Diego, which is right on the Mexican border, less than a third of residents are Hispanics, but in Los Angeles, nearly half are. In Denver, New York, and Chicago, about 30% of residents identify themselves as Hispanic, but only 5.2% in Atlanta (■ Fig. 1.6).

Suburban space has long been dominated by middle- and upper-class white Americans. This pattern has changed in recent decades. As assimilation increased, second- or third-generation members of minority groups left the core cities and moved to the surburban area, where they are not evenly distributed but concentrated in specific locations. In addition, more and more first-generation immigrants settle immediately outside the core cities, preferring locations with many jobs where a large number of members of their own ethnic group already live. Li (2009) coined the term ethnoburb for the three communities east of Los Angeles, Monterey Park (pop. 60,000), Alhambra (pop. 83,000), and Arcadia (pop. 56,000), where more than half of the residents are of Asian, especially Chinese, origin. Chinese first moved to the suburban area in large numbers after the 1965 riots in Watts, a Los Angeles neighborhood, where they were concentrated in the three places mentioned above. Since the 1980s, many new Chinese immigrants have settled here immediately. The percentage of Asians now peaks at 67% in Monterey Park (U.S.: 4.8%). *Ethnoburbs,* according to Li (2009), are multi-ethnic communities in suburban areas where

a single ethnic group strongly dominates but does not necessarily constitute half of the population. They allow members of ethnic groups that are minorities in the U.S. to thrive economically without having to conform to the values of the majority American (white, non-Hispanic) society. According to Li (2009, p. 172), unlike ghettos, ethnoburbs must have been created on a voluntary basis. According to this definition, an ethnoburb is also the city of Dearborn (2010: 98,000 inhabitants), located in Detroit MA, where the corporate headquarters and a large manufacturing plant of the Ford Motor Company are located. 38% of Dearborn's population is of Arab origin (► www.census.gov). Dearborn is home to the largest mosque in the country and is home to the Arab American National Museum, the only museum of this ethnicity in the United States. Since late 2011, the television network TLC has been airing the reality show *All American Muslim* featuring families from Dearborn. Ethnoburbs are no longer an exception, as blacks or *Hispanics* now make up by far the largest group in many suburban communities. Ethnoburbs are often created due to social and economic restructuring processes. In northern MA Miami, blacks dominate in many communities, while west of the core city, hispanics dominate (► Fig. 5.26). At least in communities with a large black population, residential location is likely due to displacement processes. According to Li's (2009) definition, these are not ethnoburbs. Whether this distinction is meaningful is another question.

Growth and Decline of the City

Contents

2

2.1 Phases of Urban Development

American cities were founded in different phases, reflected in layout and function. A rough distinction can be made between the cities located on the East Coast, which were founded in the pre-industrial period, the cities in the Midwest, which grew into large cities within a few decades due to industrialization, and the cities in the South and West of the country, some of which were founded only in the twentieth century and whose intensive growth took place in the post-Fordist phase. In recent decades, all cities have been equally affected by global restructuring processes. Some of the cities have now overcome the crisis caused by deindustrialisation, while in others the number of inhabitants is still declining and no compensation has been made for the loss of industrial jobs. The building fabric of these cities is characterised by severe decay, while in the cities mentioned first the signs of recovery are clearly visible, especially in the inner cities.

2.1.1 Pre-industrial Cities

The early cities of North America were modeled after European cities, but reflect the ideals and ideas of the different immigrant groups that settled the North American continent. The Spanish laid out cities with a square plaza and the British with a village green, which survive to this day in the center of Santa Fé or Boston. Other cities, such as New York, were given an irregular layout in the early stages that still characterizes the streetscapes south of Wall Street. In the seventeenth century, the first cities were laid out in a checkerboard layout, which soon became widespread because it was quick to implement. It also made it easier to buy or sell land to investors or future settlers who were not present. Preferred locations for early settlements were bays or estuaries on the east coast, which guaranteed access to the hinterland, or inland, the banks of rivers or lakes, which served as transport routes and suppliers of the water needed in the production process. Warehouses, mills, transhipment facilities and nearby housing for dockworkers were built on the banks, along with neighborhoods of craftsmen, tradesmen, taverns and other amusement establishments. The checkerboard layout was consistently followed even when it was actually completely inappropriate for reasons of topography. The steep hills of the City of San Francisco, with its fall-line streets, present a major challenge to pedestrians and street traffic (Cullingworth and Caves 2008, pp. 52–53; Muller 2010, pp. 304–313) (◘ Fig. 2.1).

Underlying American society is a belief in a capitalist economic system and a liberal social philosophy. A

◘ **Fig. 2.1**   San Francisco

firm belief in individual freedom, equal opportunity, and the unfettered pursuit of profit have affected the vision and reality of the North American city in many ways. Successful businessmen, wealthy property owners, and other well-off citizens have determined the destiny and public affairs of a city. It was believed that a city should be run like a business enterprise and that successful action would serve everyone. Great importance was attached to the acquisition of land, as it was ideally suited as an object of speculation. Land became a commodity and opened up the possibility of increasing personal wealth. The owners of land could decide on its use almost without restriction; exceptions were only made if public order and security were endangered. Land was most expensive in the central areas of the city, and prices fell as the distance from the center increased. Two- to four-storey rows of houses, above which only the church towers or the masts of sailing ships docked in the harbour towered, characterised the townscape. As the cities expanded, spatially differentiated uses emerged and quarters were formed. Retail, crafts, banking, or other functions were concentrated in certain parts of the city (Muller 2010, pp. 304–307; Schneider-Sliwa 2005, pp. 62, 75).

Although city life offered many advantages, an anti-urban attitude set in early on. President Thomas

Jefferson (1801–1809) described large cities as endangering the health, morals, and liberty of man (Letter to Benjamin Rush, cited in Owen 2009, p. 19). The work of natural philosopher Henry David Thoreau (1817–1862) reflects the anti-urban attitudes of Americans. A Harvard graduate, he was unable to find employment due to the recession of the time and retreated to a cabin in the woods near Concord, New Hampshire for a good 2 years. In his novel *Walden*, which is still widely read today, he glorifies the innocent life of nature, even if he doesn't recommend it for everyone. On closer inspection, however, the cabin was closer to the center of Concord than to true wilderness. The Sierra Club, founded in 1892, saw city life as poisonous to body and soul, and the National Park Service, created in 1916, wanted national parks to provide a corrective to unwholesome cities (Owen 2009, pp. 18–19; Thoreau 1854).

2.1.2 Industrial Cities

The USA entered industrialisation after the end of the Civil War (1861–1865), which had been prepared in the preceding decades by the construction of the railways (◻ Fig. 2.2). As the transport network expanded, the movement of goods was less and less tied to waterways, which therefore became less important as a location for the establishment of new towns. The Northeast was the industrial heartland of the country until the mid-twentieth century. The cities were equally the destination of immigrants from Europe and of in-migrants from the southern states, and in some cases doubled their populations within a decade. In the first half of the twentieth century, about 70% of all jobs in the secondary sector were located in the so-called *manufactur-*

◻ **Fig. 2.2** A former steelworks near Birmingham city center now serves as a museum

ing belt (Berry 1980, pp. 3–5). As larger manufacturing facilities were built, industry shifted locations along rail lines, and corridors of industrial use developed that extended far into the surrounding countryside of cities. In the early twentieth century, a mass market for automobiles, machinery, and consumer goods gradually developed, and thousands of workers flocked to factories daily. The factories were surrounded by parking lots, railroad and streetcar tracks, amusement districts, and unattractive working-class neighborhoods. The environment was not spared; industrial waste was dumped uncontrollably on heaps or dumped into rivers. Toxic emissions were released unfiltered into the air, with no regard for neighbouring residential areas where workers and immigrants lived. At the turn of the twentieth century, private organizations became increasingly concerned with improving housing, sanitation, and social and medical services. Municipalities expanded sewer systems and enacted building codes (Muller 2010, pp. 312–313, 324). Meanwhile, the wealthier population preferred residential locations far from the factories or moved to suburban areas, where housing estates were already being built near the new railway stations at the end of the nineteenth century.

The horse-drawn tram lines initially converged in the inner cities, and later the routes of the railways or subways. Large office buildings and prestigious railway stations were built here, bringing commuters from the surrounding countryside and travellers into the cities. Hotels, expensive restaurants, elegant theaters, movie houses, and cheap establishments were built for the entertainment of travelers and locals (Muller 2010, p. 309). Since the highest returns were in the center of the cities, it was important to make the best use of the land. The construction of high-rise buildings seemed almost inevitable. The invention of the (safe) elevator and steel construction were important prerequisites for the building of skyscrapers. Although Elisha Otis was not the inventor of the elevator, he had first presented an elevator with a safety brake at the New York World's Fair in 1853. The usable floor space of what was only a small plot of land could now be increased many times over and the land value increased. Chicago had largely burned down in 1871 and offered good opportunities for the construction of new tall buildings, but the demand for floor space was far greater in New York. One of the first truly tall buildings, at 119 m, supported entirely by a steel skeleton, was the Park Row Building in New York, completed in 1899 (Glaeser 2011a, p. 140). The Singer Building in Lower Manhattan, completed in 1908, was the first building in the world to exceed 150 m in height, at 186 m. Only a year later, it was succeeded as the tallest building in the world by the Metropolitan Life Tower (213 m) and in 1913 by the Woolworth Building (241 m).

2

The latter held this title until the Bank of Manhattan Building (319 m) opened in 1930, replaced only a few months later by the Chrysler Building (319 m), which in turn was edged out by the Empire State Building (381 m) in 1931. The economic crisis of the early 1930s put a temporary end to the competition for the tallest building. Within a few decades, the skyline of the North American city had changed forever (CTBUH 2011, p. 55).

The value of parks created during the period of industrialisation should not be underestimated, as the early creation of green spaces prevented later development and created green *"lungs"* in or near densely built-up centers. Many cities looked to Central Park, opened in New York in 1859 in the English landscape park style, as a model (◘ Fig. 2.3). Central Park had been designed by landscape architects Frederic Law Olmsted and Calvert Vaux, who considered urban green spaces to be of great importance for the health and social cohesion of the population. San Francisco's Golden Gate Park, opened in 1870, is similarly famous. Some cities such as Boston, Buffalo, Chicago and San Francisco created an entire system of parks linked by boulevards in the nineteenth century (Muller 2010, p. 324). A further boost in attractiveness was given to many cities as a result of the City Beautiful movement, which emerged in conjunction with the World Columbian Exposition in Chicago in 1893. The architect Daniel Burnham had built the White City with monumental buildings for the World's Fair, which burned down almost completely a few years later. At the same time, a reform movement was forming with the goal of eliminating the cities' poor housing and other ills. Burnham designed the *Chicago Plan of 1909*, which centered on large boulevards and an attractive design for the shore of Lake Michigan. Other cities, such as Cleveland, Detroit, and Washington, D.C., adopted important impulses from the Burnham Plan (Miller 1997, pp. 549–551).

2.1.3 Post-Fordist Cities

In the mid-1950s, the U.S. city entered a crisis that cannot be explained monocausally. The cities in the *manufacturing belt* were confronted with new challenges due to deindustrialization and changes in production processes. The industrial cities had been characterized by huge factories in which comparatively poorly educated workers performed monotonous tasks and produced standardized products. Adaptation to post-Fordist production processes was hardly possible under these conditions. The influence of the trade unions that had emerged in the eastern United States since the turn of the nineteenth and twentieth centuries also proved to be extremely damaging. In 1935, the *National Labor Relations Act* was passed, which made it difficult to fire striking workers and allowed for the formation of *closed shops*, i.e. factories in which all workers had to be unionized. The unions were able to enforce high wages in the textile, steel and automobile industries, which, however, led to the relocation of production to other regions. The 1947 *Taft-Hartley Act*, which allowed states to pass laws against the establishment of *closed shops*, also contributed to this. Many southern states made use of this option, creating an important precondition for the maintenance of low wages and subsequent industrial settlements (Glaeser 2011a, pp. 42, 50–51). At the same time, relocation occurred at the micro level. Downtowns in particular suffered from decentralization within urban areas. Shopping malls and office buildings sprang up in large numbers in suburban areas, while retail and other services in downtowns experienced a decline in importance. Industry also relocated to suburban areas, because with the expansion of the expressway network and in the face of falling transport costs, location on a waterway was no longer important (Berger 2007, p. 47).

In the 1950s and 1960s, the core cities suffered from race riots, as blacks were no longer willing to put up with discrimination and violence at the hands of white police officers. Many cities saw violent riots and street battles with many deaths. Businesses were vandalized and looted, and the affected neighborhoods resembled war zones. Even after the passage of the *Civil Rights Act* in 1964, which prohibited discrimination against racial, ethnic, religious, or other minorities, the riots were slow to subside at first. Cities evolved into seemingly lawless and extremely dangerous spaces, and politicians had contributed to this. Already at the beginning of the twentieth century, the misguided policies of many cit-

◘ **Fig. 2.3** New York's Central Park enjoys great popularity at any time of year

ies, the venality of politicians and police, and the inadequate fight against crime had been lamented (Steffens 1904). The murder rate had fallen sharply in the United States from the 1930s until 1960, when it began to rise again. In New York City, it quadrupled between 1960 and 1975, when it reached a sad record of 22 murders per 1,00,000 residents (Glaeser 2011a, p. 108). At the same time, social unrest and poor policies exacerbated urban decline. The crisis reached a peak when New York could no longer pay the wages of city employees in 1975. At the last moment, the federal government pledged financial assistance, averting the city's financial bankruptcy. Images of mountains of garbage in the streets of New York were broadcast worldwide. The North American city had reached a globally visible low.

Core cities lost more and more jobs, and tax revenues fell as the middle and upper classes and businesses moved to suburban areas. The percentage of poor and minorities increased, housing stock aged, infrastructure deteriorated, and public services were reduced, especially affecting schools. At the same time, the construction of museums, hotels, and convention centers, the restoration of historic buildings, and an increasing demand for housing in the inner cities heralded a turnaround as early as the beginning of the 1980s, but at the time one could only speculate about its sustainability (Conzen 1983, p. 149).

2.1.4 Postmodern Cities

While the cities of the manufacturing belt were in a downward spiral from the 1950s onwards, many cities in the south and west of the country experienced an immense increase in population and significance in the second half of the twentieth century. This was triggered by the establishment of new industries, investments in the military and aerospace, and the influx of people from the Northeast or from Mexico and Central America. At the same time, some of the old industrialized cities succeeded in switching to knowledge-intensive services, while other cities were unable to halt their decline.

Influenced by global restructuring processes, the US city has evolved from an industrial city to a postmodern city. The industrial age was characterized by largely uniform wages, and career advancement was almost exclusively possible within a company. Businesses competed with local or, at best, national businesses, while competitors in post-industrial society can be located anywhere in the world. New forms of communication, falling freight rates and the liberalisation of world trade enabled the emergence of global value chains. It is not uncommon today for the components of a finished product to be manufactured in a number of countries. The worker in Chicago no longer competes with workers in neighboring Gary, New York, or Baltimore, but with workers in China, India, or Vietnam. As demands have risen, wages for high-skilled workers have risen; but fewer jobs are available for low-skilled workers in the industry (Clark et al. 2002, p. 504). The decline has been offset by an increase in jobs in services, where there is also a juxtaposition of workers with very high wages, such as those in finance, and those with precarious, low-wage jobs, such as those in the fast-food industry or the hotel industry.

While the industrial city was tied to convenient locations and good natural conditions, in the postmodern era places with unfavorable conditions that had only a few thousand inhabitants a few decades ago have risen to become large cities with hundreds of thousands of inhabitants, as the developments of the desert cities of Phoenix or Las Vegas show. Some of the up-and-coming cities, such as San Jose in Silicon Valley, hardly have a recognizable center anymore. But cities in the *manufacturing belt* have also experienced a decentralization of all functions. At the same time, processes such as *gentrification*, privatization, and globalization have led to lasting changes. The postmodern city is a chaotic city that seeks to achieve global attention through spectacular events or architecture. It is run like a business enterprise that seeks to attract a global elite, global capital and international tourists. Services, and in particular telecommunications and finance, are the leading economic sectors in cities where consumption is more important than production. Since not all residents benefit from this development, the postmodern city is a fragmented and polarized city, composed of a large number of isolated neighborhoods, similar to a patchwork (Gratz and Mintz 1998, pp. 33–34).

2.2 The Suburban Space

The suburban area consists of the politically independent municipalities and municipality-free areas that are administered by the counties. In addition, there are enclaves that lie within the core cities and are counted as suburbs by definition, but are often not perceived as such. Examples include Beverly Hills and Highland Park, which are located within the boundaries of Los Angeles and Detroit, respectively.

Since the 1950s, three-quarters of the population growth in the United States has been concentrated in suburban areas, where half of all Americans lived at the turn of the millennium (Kotkin 2001, p. 1). Given these rates of growth, the suburbs represent, at first glance, an extraordinary success story. However, they are now as different as the core cities and have adapted to them in many ways. Many of the problems of core cities also

2

exist in suburban areas, and suburbanization is increasingly viewed critically. On the one hand, the suburbs embody ideals in the sense of an Arcadian utopia; on the other hand, they are the result of an endless supply of cheap building land, cheap transport, and lax specifications by planning and administration (Knox 2005, p. 33).

The form and function of the suburbs can only be understood from their historical development. Compact cities, in which commerce, trade, and dwellings crowded into a small area, existed in the USA only for a short time and almost exclusively in the early settled northeast of the country. Faced with poor housing conditions, high population density, dirt, and often stench in cities, many immigrants developed anti-urban sentiments early on (Muller 2010, p. 319). Technological advances had already begun to transform the American city between 1815 and 1875. The introduction of steam-powered ferries, horse-drawn buses, commuter trains, and cable railways allowed the city to expand over a wide area. With the expansion of the railroads, which created the opportunity for daily train travel to work, numerous new suburbs sprang up on the periphery of cities in the second half of the nineteenth century. Particularly in the environs of the rapidly growing cities of Chicago and New York, streetcar suburbs were created along the radial rail lines built from the major cities into the surrounding countryside. Since a walk of more than 1.6 km to the next stop was no longer considered reasonable, suburbs were preferably built near the railway stations (Knox and McCarthy 2012, pp. 67–68; Jackson 1985, pp. 20, 92–97).

The early suburbs represented a counter-model to the industrial city, which was afflicted with many ills, and were idealized retreats for the affluent. At the same time, suburban space developed into a consumer good whose value, like that of any other commodity, was determined by supply and demand (Hanlon 2010, p. 15; Knox 2005, p. 35). As early as the nineteenth century, the practice of subdividing the available land into subdivisions of varying sizes and selling them to investors who planned and laid out the new settlements as developers took hold (Cullingworth and Caves 2008, p. 118; Jackson 1985, p. 135). Soon the checkerboard layout was deviated from, as it was considered unattractive in suburban areas if wealthy buyers were to be attracted. In the Orange Mountains of New Jersey, beginning in 1852, developer Llewellyn S. Haskell purchased land with a spectacular view of New York to have landscape architect Alexander J. Davis lay out an estate designed to reflect the natural beauty of the landscape for wealthy New Yorkers. The gently curving streets traced the natural landscape and gave the estate a pastoral character. For the first time, a house pattern book was published,

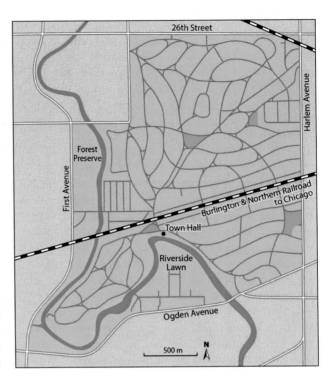

▫ Fig. 2.4   Riverside Street Map. (Rand McNally 1998, Adapted from CyberToast Web Design and Hosting)

detailing the style of the houses to be built, as a guide for potential builders. Houses in the style of Italian villas or Swiss chalets, for example, were envisaged. More famous than Llewellyn Park, however, was the suburb of Riverside, planned in 1868 by the well-known landscape architect Frederick Law Olmstead 16 km west of downtown Chicago (▫ Fig. 2.4). Riverside was the first stop on a new railroad in the suburban area. For Olmsted, the suburbs were not refuges to shelter from the ills of cities, but a synthesis of city and wilderness. On the periphery of the cities, free-standing houses were to be built embedded in a natural environment, and at the same time residents were to be able to enjoy the benefits of the city. Park-like Riverside, with its generous lots and villa-like homes, is considered the first true master planned community in the United States (Knox and McCarthy 2012, pp. 78–81; Jackson 1985, pp. 76–79; Chicago Historical Society 2004, keyword: Riverside, IL).

2.2.1 Standardisation

The concept of the master planned community, in which a developer designs the layout and land use of a neighbourhood or an entire settlement, became established in the development of suburban areas in the following decades (▫ Fig. 2.5). It quickly became apparent

Fig. 2.5 Construction of a *master planned community* in Las Vegas

Fig. 2.6 Construction of a *balloon frame house* in San Diego

that there was as much to be made from the acquisition of building land, the development of settlements, and the subsequent sale of individual developed or undeveloped lots as from the production of manufactured goods or the commodity trade. From 1880 to 1882, Samuel Eberly Gross of Chicago developed 40,000 lots in 150 subdivisions, laid out 16 towns, and built and sold more than 7000 homes. More typically, however, a developer worked with a larger number of subcontractors. Sam Bass Warner had subcontractors build about 25,000 homes in Roxbury, West Roxbury, and Dorchester in the Boston area between 1870 and 1900, but none of them built more than 3% of all homes. In Kansas City, Jesse Clyde Nichols built about 10% of all homes built during that period between 1906 and 1953. He also laid out County Club District, a 26 km^2 suburb southwest of the city for about 35,000 residents. The new community was laid out in the style of the European garden city idea and the American City Beautiful movement with many open spaces (Jackson 1985, pp. 135–136, 177).

The rapid construction of the new suburbs would not have been possible without the *balloon frame house*, in which a wooden framework consisting of thin battens is first erected and then the walls are hung in the framework (**Fig. 2.6**). The construction of these houses is cheap, quick, and no great skill is required. The method of construction was probably used sporadically in colonial times, but did not catch on until Chicago, founded in 1833, from where it quickly made its triumphant march across America. While only the wealthy could afford the expensive stone houses, the simpler wooden houses were also affordable for the middle class (Jackson 1985, pp. 124–126; Chicago Historical Society 2004, keyword: *balloon frame house*).

2.2.2 State Support

After the turn of the twentieth century, technological progress, the expansion of the road network and mass transit, and soon government programs spurred suburbanization. In 1913, Henry Ford introduced assembly line production in Detroit for the construction of his *Model T* and reduced production costs many times over. The private automobile evolved into a means of mass transport, providing individual transport to anywhere in the urban hinterland and opening up the land between rail lines. By 1908, New York State had opened the first part of its Parkway system to provide quick access to the city of New York for residents of the surrounding areas. Across the country, the federal government provided organizational and financial support for the construction of paved roads through the passage of the *Federal Road Act* of 1916 and the *Federal Highway Act* of 1921, which allocated $75 million for the task (Glaeser 2011a, p. 173). The introduction of the telephone into private homes made it possible to communicate with relatives or friends in other locations at any time (Kotkin 2001, p. 3). There was no stopping the uncontrolled urban *sprawl*. Even the middle class could now realize the dream of seeing their children grow up in a rural idyll (Muller 2010, p. 319). At the same time, the complete disenchantment of the last remaining utopian ideals set in. City dwellers moved into country houses in suburban areas without wanting to give up their accustomed urban life. Rarely, respected architects and planners developed new house forms or floor plans. Increasingly, new suburbs became copies of existing suburbs, whose external appearance was determined by the financial means of its inhabitants. The wealthy were able to retire to large mansions along curved and tree-lined avenues, while the middle class had to be content with small houses and plots of land in estates with checkerboard layouts. (Almost) identical suburbs emerged in different locations, and the former utopias became faceless suptopias (Knox 2005, p. 35; Rybczynski 1996, pp. 173–196).

Towards the end of the Great War, the first government programs to promote the purchase of owner-occupied homes were adopted, but they were not very successful. Major action was not needed until the American economy and construction industry collapsed in the wake of the 1929 banking crash. Between 1928 and 1933, home construction fell 95%. At the same time, the sale price of homes dropped. More and more citizens were unable to service their loans, and foreclosures increased. In 1933, President Roosevelt (1933–1945) signed legislation to protect property owners from foreclosures, and a year later the Federal Housing Administration (FHA) was created with the goal of reducing high unemployment and boosting home construction. The FHA introduced mortgage protection mechanisms that reduced the required down payment to 10% of the home price from about 50% previously. It also extended the terms of mortgages. Monthly payments became lower, and the number of foreclosures decreased. Since rates were often lower than rents, it was cheaper to build a home than to rent. A mass movement of whites into suburban areas began, because only they had access to the cheap mortgages. The FHA thus encouraged not only suburbanization and the emptying of core cities, but also ethnic segregation. Only in a few exceptional cases had the first suburbs been built around 1900 for industrial workers and for residents with black skin. In 1944, the government passed the *Servicemen's Readjustment Act*, known as the *G. I. Bill of Rights*, to allow 16 million returning servicemen to purchase a home without a down payment (Beauregard 2006, pp. 78–80, 107; Hanlon 2010, pp. 3–15; Jackson 1985, pp. 192–206). At the same time, the distance between home and work increased. Moreover, the private car made it possible to have a job in another suburb. By 1934, one in eight heads of households commuted to another suburb rather than to the core city (Jackson 1985, p. 182). The automobile undoubtedly increased the owners' range of movement. A disadvantage, however, was that private vehicles required more and more space for roads and for parking (Glaeser 2011a, p. 177).

2.2.3 Increasing Suburbanisation

In the first half of the twentieth century, the single-family home clearly emerged as the preferred form of housing. Of the 6 million housing units built between 1922 and

◘ **Fig. 2.7** Typical residential house in a suburban area

1929, more than half were single-family homes, most of which, again, were built in suburban areas (Rybczynski 1996, p. 175) (◘ Fig. 2.7). During this period, suburbs have grown twice as fast as core cities (Kotkin 2001, p. 3). Nowhere does the American Dream, first coined and defined by Truslow Adams in 1931, seems better embodied than in the suburbs, where everyone hopes to live in their own house on a large lot in a happy family. The American Dream has become deeply engrained in the American consciousness and embodies the quest for a better and more equitable life where everyone is equal and all have identical opportunities for advancement (Beauregard 2006, p. 142; Hanlon 2010, p. 1).

In 1926, the newly established Advisory Committee on City Planning and Zoning in Washington, D.C., passed a law that regulated the planning sovereignty and competencies of municipalities in land use designation (*zoning*) as well as the preparation of master plans, which are comparable to German land use and development plans. Based on this framework law, the individual states passed their own laws for the adoption of master plans in the following years. *Zoning* emerged as the most effective tool for suburban development. Local politicians and planners calculate the ratio of residential building lots to other uses and can thus reliably calculate future tax revenues. The size of building lots continues to influence the future social structure of communities to this day. Large plots of land guaranteed the construction of large and expensive houses and high revenues from property taxes, which in the USA, along with the sales tax, account for a large share of municipal tax revenues (Freund 2007, p. 217).

During World War II, tax depreciation was increased for building homes, but not for rehabilitating them. It was now often far cheaper to build a new house than to repair an old one. The idyll of the countryside was rediscovered, for it promised a safe life away from the dirt and dangers of the big city. From 1940 to 1980, home ownership rose from about 40% to two-thirds; since then it has been stable. New homes were preferentially built on the periphery, where land prices were low. The growth of small communities and the retreat to rural areas on the outer periphery of metropolitan areas is known as counter-urbanisation. Some of the new housing development has even occurred in exurbia, that is, in a purely agricultural setting without urban amenities. The abandoned houses in the core cities were occupied by lower income earners and immigrants. Because the number of new housing units in many years far exceeded the increase in population, houses that were in very poor condition were not reoccupied and entire neighborhoods were abandoned (Berry 1980; Conzen 1983, p. 143; Kotkin 2001, p. 4).

After World War II, mass consumption took hold at all levels in the USA. More and more people identified themselves through consumer items such as clothes and cars, and lifestyles manifested themselves in certain consumption patterns. Nothing was more appropriate for the (perceived) enhancement of one's status than living in an expensive house in an affluent neighborhood in the suburban area (Knox 2005, p. 36). Increasing standardization in the layout of new developments and house construction lowered costs and made owning a house in the suburban area a mass consumer good like any other. No investor was as successful in reducing construction costs through standardization as William Levitt of New York. For example, because Levitt did not hire union workers, he did not have to have the houses painted manually, as the unions demanded, but could use spray paint. Between 1947 and 1951, Levittown was built on Long Island in New York State on an area of 18 km^2 with nearly 17,500 single-family homes for 82,000 residents that were affordable to the lower middle class at a price of less than $8000 ($65,000 in 2009 prices). Architectural critics complained about the monotonous design, but at the time the houses were considered comfortable. In the intervening decades, most Levittown residents have remodeled or added on to their homes, and in 2007, when the development celebrated its 60th birthday, hardly any homes here sold for less than $350,000, making them unaffordable to the original target demographic (Bruegmann 2005, p. 43; Hanlon 2010, pp. 2–4; Short 2007, pp. 34–35).

Encouraged by the further expansion of the road network, government subsidy programs, improved tax depreciation for investments in suburban areas, high crime rates and racial unrest, and poor schools in the core cities, a mass movement into suburban areas began in the 1950s. The *Civil Rights Act*, passed in 1964, secured equal rights for all minorities as for whites and encouraged suburbanization. To ensure equal opportunity for children

of all ethnicities, and especially to break the intellectual isolation of black ghettos, students were bused to schools outside their own neighborhoods. Because many parents feared that black students would lower the standards of the schools, and because transport was not allowed across school district lines, many whites moved to suburban areas. This trend came to be known as *white flight*. It was not until the 1980s that members of ethnic minorities began to move, at first hesitantly and then with increasing frequency, from the core cities to the suburban areas. At the same time, new immigrants increasingly chose the suburbs as their first place of residence (Fishman 1987, p. 183; Glaeser 2011a, p. 90).

The first *homeowner associations* (HOAs) were founded as early as the nineteenth century and became widely accepted in the 1960s. All subsequent residents of a master planned community must join HOAs even before purchasing the lots or properties. After completion, the developments are managed by the owners themselves as *common interest developments* (CIDs), and some of these neighborhoods are set up as gated communities with access via a single road. Similar to the declarations of partition and community regulations in apartment buildings with condominiums known in Germany, the CIDs regulate the ownership and financing of the common property as well as the rights and obligations of the residents in a binding manner. However, they are much more far-reaching than is usually the case in Germany. HOAs may stipulate that no cars may be parked in driveways, garage doors must be closed at all times, all curtains must be blue and flowers in the front garden must bloom red, that hanging laundry in the garden is prohibited and outdoor play equipment must be made of wood. In addition, they often independently organize garbage collection or street cleaning, as well as neighborhood surveillance, thus relieving the municipal coffers. A master planned community is planned for a specific target group or more precise income bracket. Wealthy buyers are attracted by large lots with large houses, while subdivisions for the less financially strong buyers are divided into many small lots. In extreme cases, to save costs, only one type of house is built in a neighborhood. Since 1948, it has not been possible to exclude members of certain ethnic groups, but a minimum age for residents may be set, which is an important basis for building senior housing developments. HOAs and CIDs have encouraged segregation and fragmentation of the American urban landscape. Paradoxically, Americans see freedom as their highest good, but at the same time voluntarily submit to strict rules of coexistence, thus overriding the fundamental rights guaranteed by the Constitution and later amendments (Lichtenberger 1999, p. 32 f.; McKenzie 1994, pp. 18–38; Rybczynski 1996, p. 182).

2

According to his own information, the Texan D. R. Horten is now the largest developer of residential real estate in the USA. Horton has built neighborhoods in 26 states and completed nearly 19,000 homes in 2010 alone (▶ www.DRHorton.com). Master planned communities that have attracted particular attention in recent years include Celebration, built by the Disney Company in Florida, and Ave Maria, also built in Florida by the founder of Domino's Pizza. The focal point of Ave Maria is a Catholic church around which all other buildings, including a Catholic university, were built. Ave Maria is meant to provide a life based on Catholic values. Since it is not possible in the USA to exclude people on the basis of their religious affiliation, however, people of other faiths may not be denied entry if they really want to (▶ www.avemaria.com).

2.2.4 Suburban Worlds and Lifestyle

In recent decades, the original idea of utopian life in the suburban space has been resurrected in a new form. Idealized worlds are simulated and not infrequently realized behind walls with the simultaneous complete regulation of all areas of life. People are only connected to each other through contracts, otherwise there are no social relations with neighbours and there is a lack of any kind of responsibility for the community (Knox 2005, p. 37). Names like "*King Farm*" evoke the association of living in nature, even if the farm was destroyed for the construction of the exchangeable neighborhood. In King Farm not far from Washington, D.C., all the naturalistic elements such as small streams are tamed, evoking English country parks rather than wild nature. Residences harken back to historic models and are built in colonial revival or Tudor styles. The names of master planned communities, like those of their subdivisions, are primarily for marketing purposes (Gerhard and Warnke 2007). Master planned communities often have names such as Vineyards Fields, Old Farm Road, or Heritage Hills, which suggest a connection to historic places that usually never existed at the site in question (Gratz and Mintz 1998, p. 139). The oversized apartment buildings in many developments are built for the well-funded upwardly mobile of the new economy, who take it for granted that they will live in neighborhoods with every conceivable amenity while banishing all unpleasantness such as crime, decay, and traffic. Single-family homes built in 2010 had an average living area of 222 m². Nearly 40% of the homes had at least four bedrooms, although families are getting smaller in the U.S. as well (▶ www.census.gov). Knox (2005) refers to the houses of the suburban area with their huge entrance halls and oversized rooms as vulgar and the

neighborhoods in which these houses are concentrated as *vulgaria*, while Kotkin (2001, p. 144) prefers the term *nerdistan* in reference to the nerds (socially isolated computer enthusiasts) living here.

As the distance from the core city increases, land prices decrease. One consequence is that towards the outer periphery the individual parcels become larger and larger (Jackson 1985, p. 7). Metropolitan areas have grown in area over decades, and there is little doubt that this process will continue. Often, it is simply a matter of holding out and waiting until residential development has advanced to their own land and land values increase. Mike Davis (1992a, pp. 153–164) provides an impressive description of the rise in land and property prices in only a short period of time during the development of Orange County in Los Angeles MA. According to Davis, the owners of the building land and the homeowner associations are the real rulers over a region, because they determine prices, the use of the land, and to whom it is sold. The suburbanization of the outer periphery is given additional momentum by the constant expansion of the expressway network. According to Baum-Snow (2010, p. 8), each new highway built from the center of a metropolitan area into the suburban area results in an 18% loss of population in the core city. The further one moves from the center toward the periphery, the greater the dependence on the automobile. Wide streets often lacked any kind of pedestrian crossings, and housing developments lacked sidewalks. Stores and shopping centers are surrounded by dozens, hundreds, or even thousands of parking spaces, while restaurants, banks, pharmacies, or dry cleaners offer drive-in service (Muller 2010, p. 323). Walking is not provided for in suburban areas. Adults drive, and children are driven to friends or sports by soccer mums and to schools by buses.

2.2.5 Heterogeneous Suburbs

On the one hand, most master planned communities today are completely faceless and interchangeable; on the other hand, the suburbs have become increasingly heterogeneous. Some suburbs are heavily industrialized, while others are almost exclusively residential. But even these suburbs can differ greatly. In quite a few suburbs, such as north of Chicago or in Westchester County west of New York City, the super-rich are concentrated. Other suburbs are marked by job losses, population drift, and deteriorating housing stock. This is true, for example, of Camden, which faces Philadelphia across the Delaware River, and East St. Louis, which faces the city of St. Louis on the east bank of the Mississippi River. These two communities have been among the poorest in the United States for many years. Suburbs have long ceased to be synonymous with social advancement and success, and since the turn of the millennium, the number of poor people in suburban areas has grown faster than in core cities. More poor people now live in the suburbs of large cities than in the cities themselves. The problem is that there are hardly any social institutions in suburban areas that can provide help in an emergency. The poor here rely far more on themselves than in the large cities, which have a long history of serving an indigent population (Kneebone and Carr 2010; Hanlon 2010, pp. 12–17). For decades, the population of suburbs, like household and family structures, has become increasingly heterogeneous. The proportion of households with white middle- and upper-class families has declined in favour of members of ethnic minorities (Gober 1989, p. 311), and increasing numbers of singles and older people live in suburban areas. In many ways, the trend in suburban areas is similar to that in core cities. Lang et al. (2009, p. 727) refer to this process as quasi-urbanisation.

The population development of many suburbs has adjusted to that of the core cities, because in the meantime the suburbs are also losing inhabitants, especially in the old industrialized regions. According to a study conducted in 35 metropolitan areas, the population declined in one-third of all suburbs in the 1990s. In Pittsburgh MA, 108 out of 128 suburbs experienced losses; similar was true for suburbs surrounding Buffalo, Philadelphia, Detroit, and Cleveland (Lucy and Phillips 2000, pp. 166–173). Most inner-ring suburbs, created during the first suburbanization phase and directly adjacent to core cities, have developed differently than outer-ring suburbs, which have emerged similar to the growth rings of a tree on the outer periphery of metropolitan areas (Hanlon 2010, p. 9; Short 2007, pp. 41–42). Many of the inner-ring suburbs have been plagued by decay and population loss for decades. The original residents have either moved to higher-status neighborhoods or died, and the remaining population has a comparatively high median age (Berry 1980; Jackson 1985, p. 301). Many of the mostly older homes in the inner-ring suburbs are no longer up to current standards or are in need of rehabilitation. Because homeowners, investors, and the public sector prefer to build new, abandoned houses cannot find buyers or tenants (Lucy and Phillips 2000, p. 2; Hanlon 2010, p. 54). On the downside, most of the older suburbs with blight are not eligible for public assistance. Funds from the CDBG (*community development block grant*) program are only paid to low-income neighborhoods in cities with at least 50,000 residents or in selected urbanized counties with at least 200,000 residents (Hudnut 2003, pp. 85–90). HOPE VI, which is part of the *U.S. Housing and Development Program*, promotes only the conversion of homes built with public funds into mixed-use developments, but these are rarely located outside of the major core cities. Alexandria near Washington, D.C., Camden near Philadelphia, and Richmond in San Francisco Bay are among the few inner-ring suburbs that have received HOPE VI funds (Hudnut 2003, pp. 259–260). Inner-ring suburbs in the metropolitan areas of the Northeast and Midwest of the United States, where industry had settled early on, are particularly affected by signs of decay. When the factories closed, the suburbs lost their economic base and many people left the communities. These include suburbs near Chicago, Philadelphia, Cleveland, St. Louis, Indianapolis, or Dallas. In many cases, poor immigrants of non-Caucasian origin have settled there in more recent times, taking advantage of the low purchase price for the mostly simple homes on small lots (Hanlon 2010, p. 27; Kotkin 2001, p. 11). However, there are also suburbs close to the core of the city that have seen positive demographic and economic development in more recent times. Some suburbs, such as Burbanks in the Los Angeles metropolitan area, have succeeded in reinventing themselves as high-tech locations, while Hemstead north of New York City and Bellaire near Houston owe their boom to the large influx of Hispanics and Asians. Based on the price development of residential and office real estate, Kotkin (2001, pp. 13–22) was also able to demonstrate a positive development for part of the inner-ring suburbs of the metropolitan areas of San Jose, Houston, Los Angeles and Atlanta.

Hanlon (2010, pp. 115–131) has examined the building fabric and socio-economic structures of nearly 1800 inner-ring suburbs and assigned them to four categories: "elite" (12%), "middle class" (35%), "precarious" (vulnerable) (47%) and "minorities" (7%). The elite suburbs were predominantly created before 1939, but in some cases as early as the end of the nineteenth century. Well-known examples are Llewellyn Park in New Jersey or Olmos Park in San Antonio. In these early elite suburbs, wealthy people built prestigious houses on large plots of land. Due to their good building fabric and location, they are still sought-after residential locations today. The income of the residents is far above average, and the poverty rate is low. In the East and Midwest, almost exclusively whites live in the inner-ring elite suburbs, while in the West some Asians and in the South and especially in southern Florida Hispanics also live here. The income of the residents of the middle-class suburbs, 80% of whom are white, is about 10% higher than the average income of all American suburbs. To the east and southeast, many originated in the nineteenth century as streetcar suburbs (see above). Middle-class suburbs in the western and southern United States are more ethnically heterogeneous than in other parts of the country. In the precarious suburbs, residents' incomes are 22% below the average earned in all suburbs, and the poverty rate is about 10%. What is sensitive is that these values have developed negatively in recent years. At the same time, the proportion of ethnic minority residents is steadily increasing. The once predominantly white middle class withdrew from some of today's precarious suburbs as early as the 1960s, when the first blacks moved in. Many of the precarious inner-ring suburbs emerged in the course of deindustrialization and are therefore often found in the Northeast in the environs of the cities of Chicago, Pittsburgh, Baltimore, Philadelphia, Detroit, Boston, St. Louis and Cleveland, but also near Los Angeles. Outside the cities of Baltimore and Chicago, huge steel mills had been built early on. Workers lived in homogeneous suburbs close to the mills, which were stable due to high wages enforced by unions in heavy industry until deindustrialization started a downward spiral. The average income of the ethnic inner-ring suburbs is even lower than of the suburbs classified as precarious, and the poverty rate is far higher at 18%. In these suburbs, which are particularly common in the western United States, there has been a widespread replacement of the population. The formerly white middle class has been replaced by poor Hispanic immigrants.

2.2.6 Industry and Service Providers in Suburban Areas

The function of suburban space has changed since the middle of the twentieth century. While in the 1950s the suburbs were still perceived as *"faceless, homogeneous bedroom communities"* (Gober 1989, p. 311), in the following decades services and industry increasingly settled here. The expansion of the road network allowed industry to relocate to suburban areas. In addition, trucks could reliably transport increasingly larger volumes and weights. As just-in-time production increased, good access to the interurban road network became increasingly important. Industry was no longer tied to locations along waterways or rail lines and moved to suburban areas, where land prices were also far cheaper than in the core city and larger reserves of building land were available. At the same time, large warehouses moved to the outskirts of the city (Jackson 1985, p. 183; Kotkin 2001, p. 5). Investment in suburban areas was also supported by improved depreciation options beginning in 1954. Whereas previously only straight-line depreciation was available, a new law allowed accelerated depreciation in the first few years after completion. The taxes that an investor had to pay on the profits from the use of a building were usually offset by depreciation. Moreover, if the

◻ Fig. 2.8 Space-consuming shopping center in Greater Miami

amount of depreciation that could be taken exceeded the amount of income, which was common, an offset could be taken against the investor's other income. This created a tax loophole of great value to potential investors (Hanchett 1996; Ture 1967).

The pioneers of the relocation of services to suburban areas were the shopping centers (◻ Fig. 2.8). The first shopping center was built in Kansas City in the 1920s, but was rarely copied in the following decades. Beginning in the early 1950s, large shopping centers suddenly opened in a variety of locations in the suburban areas of American cities. The increasing availability of the automobile and the simultaneous expansion of the highway network, especially after the passage of the *Interstate Highway Act of* 1956, are considered important prerequisites for shopping centers to establish themselves in the market. In addition, purchasing power increasingly shifted to suburban areas. The department stores were traditionally located in the inner cities, but recognized that these locations were endangered as more and more people moved to suburban areas and the accessibility of the inner cities was not optimal with a simultaneous lack of parking space. In part, the department stores initiated the construction of shopping centers, in which they took on the important role of customer magnets. Since the shopping centers required a lot of building land, investors often acquired plots of land on the outermost periphery even before housing estates had been built there. Preferred locations were intersections of supraregional roads, which developed into growth magnets (Conzen 2010, pp. 426–427; Hahn 2002, pp. 30–35).

At the beginning of the twentieth century, office buildings were found almost exclusively in the inner cities, but already in the coming decades a competing second office location developed in some cities only a few kilometers away, and secondary downtowns were formed. Midtown Manhattan, the Wilshire District in Los Angeles or Midtown Atlanta are examples of this development. Meanwhile, the older and newer office locations have often grown together and are perceived as a single downtown (Lang 2003, p. 8). Office buildings were constructed in close proximity to shopping centers beginning in the 1970s, providing increasing competition to downtown office locations. Although the prestigious headquarters often remained in the city centers, the simple activities were frequently relocated to suburban areas. Investors built either individual office buildings or, on larger subdivisions, entire office parcs with several office buildings, lawns, and numerous parking lots (Muller 2010, pp. 310–312). As the number of office jobs has increased sharply in recent decades, a great deal of attention is paid to the locations of offices because they are an important indicator of the growth of regions. At the same time, the relocation of office sites informs about urban restructuring processes. New office locations also suggest increasing urban *sprawl*. When new office parcs are built on the periphery, new neighborhoods will soon be built in the surrounding area, and traffic flows will intensify in the direction of settlement expansion in the future (Lang 2003, p. 6; Lang et al. 2009, pp. 729–730). The problem is that the new centers in suburban areas do not belong to any political unit with clearly defined boundaries, but can extend across several municipalities, if not counties.

2.2.7 *Urban Villages, Edge Cities, Edgeless Cities* and Aerotropolis

Leinberger and Lockwood (1986) studied the new concentrations of services in suburban areas and called them *urban villages*, but defined them only vaguely by high-rise office buildings visible from afar, a large daily population and a high volume of traffic. They also attested to each *urban village* having its own catchment area with a radius of about 16 km. At almost the same time, Fishman (1987, pp. 182–197) and Hartshorn and Muller (1989) noted a similar development and referred to the new centers as *technoburbs* and *suburban downtowns*, respectively. Other terms for the stand-alone centers that compete with the historic downtowns include *outer cities, edge cities, technolopolis, technoburbs, silicon landscapes, postsuburbia,* or *metroplex*. Soja (1996, p. 238) refers to them as *expopolis*, but stresses that they are urban entities, but not real cities.

The new suburban centers did not receive great attention until the 1991 book *Edge Cities. Life on the New Frontier* by journalist Joel Garreau. Garreau defines *edge cities as* follows:

- At least 5 million sq. ft. (4,64,515 m²) of office space; this is where the workplaces of the future will be created.
- At least 600,000 sq. ft. (55,741 m²) of retail space. The retail offering is equivalent to that of a large shopping center with three department stores and around 80–100 smaller shops.
- More workplaces than bedrooms and therefore a higher daytime than night-time population
- *Edge cities* are perceived by the population as a unit.
- *Edge cities* had no characteristics of a city 30 years ago. Thirty years ago, there was only a residential function here, or even just meadows.

Edge cities are preferably found at the intersections of highways. They also usually have one or more high-rise buildings that are visible from a distance. Well-known *edge cities* include Tyson's Corner in Washington, D.C. MA, Schaumburg in Chicago MA, and King of Prussia in Philadelphia MA. Garreau's remarks, however, did not stand up to empirical scrutiny. He had first made his observations in the northeastern U.S. state of New Jersey, which is separated from New York City only by the Hudson River, and here he found nine fully developed *edge cities* and two other almost fully developed *edge cities*. Based on his observations, Garreau declared the development in the New York metropolitan area to be the new prototype for the future development of polycentric regions. Robert Lang (2003) could not find a single concentration of urban functions in New Jersey in the 1990s that met Garreau's definition of an *edge city*. Lang located only three clearly identifiable concentrations of office space at the defined scale in New Jersey; however, each was several miles to the nearest large shopping center. He reached similar conclusions in Los Angeles MA, where Garreau (1991, pp. 431–432) had identified 24 *edge cities*, of which 16 were fully developed and eight were well on their way to becoming full *edge cities*. Lang (2003, p. 136) has been able to demonstrate only six *edge cities* in the region.

In the following years, Robert Lang extended his research to 13 *metropolitan areas* (Atlanta, Boston, Chicago, Dallas, Detroit, Denver, Houston, Los Angeles, Miami, New York, Philadelphia, San Francisco and Washington, D.C.) and came to the conclusion that there were far fewer *edge cities* than suggested by Garreau, and instead many smaller office locations were located more or less contiguously in a large number of locations outside the downtowns. Lang introduced the term "*edgeless cities*" to describe this phenomenon (Lang 2003; Lang et al. 2009). Because *edgeless cities* are only vaguely defined, they are often difficult to locate in space and their boundaries are hard to define precisely. *Edgeless cities* are constantly changing their shape,

2

which is why Lang calls them *elusive cities*. *Edgeless cities* are defined less by their external form than by their function, i.e. the large number of office workplaces. In the *metropolitan areas* studied by Lang, offices were predominantly located in city centers or *edgeless cities*. In 11 of the 13 case studies, the *edgeless cities* actually accounted for more office space than the central areas of the core cities. Only in New York and Chicago were the downtowns still the most important office locations in the region. Overall, only a quarter of the office space in all MAs surveyed is located in *edge cities*. Many, however, lack the retail space required by Garreau. In fact, taking into account all of the criteria that Garreau believes an edge *city* should have, only about one-fifth of the office space in the MAs studied is in *edge cities*. When Garreau made his observations in the mid-1980s, *edge cities* were growing particularly rapidly. Since the 1990s, however, most new office space has been in *edgeless cities*. Unlike the old downtowns and *edge cities*, retail and office locations are no longer identical. In *sprawling* urban landscapes, everyone drives their cars to work and shop. Since there is no longer a reason for concentrations of locations, retailers are locating where they can best reach their customers. *Edgeless cities* grow without a specific pattern and seemingly without a plan. Traditional forms such as growth in concentric rings no longer exist. The US city as it has been known for generations is dissolving. Since there is no longer a compelling spatial relationship between retail and offices, Lang has limited his research on *edgeless cities* to office locations. In 2005, Miami MA had the highest share of total office space in *edgeless cities* at 72.1%, ahead of Philadelphia MA (54.3%) and Detroit MA (54.1%). In contrast, in the New York and Chicago MAs, 54.5% and 49.2% of office space was still in downtowns, but at the same time 32.3% and 39.4%, respectively, was in *edgeless cities* (Lang et al. 2009, p. 736).

Unlike the *edgeless city*, the *aerotropolis* concept developed by John Kasarda has a center. Cities and regions have long since been connected not by rail lines, but by air routes. In the past, cities developed around railway stations; in the future, according to Kasarda, airports will form the centres around which all other functions are grouped. Highways will no longer star to the old downtowns, but to the terminals. In a first step, airports will be built far outside cities in suburban areas, in a second step the city will follow the airport, and in a third step the airports themselves will change into cities (Kasarda and Lindsay 2011, pp. 20, 189). In the USA, there are already numerous examples that illustrate the development boost an airport can give to a region. In 1941, the capital Washington, D.C., received its first airport, the National Airport (today Ronald Reagan Washington National Airport), on the banks of the Potomac River and not far from the Pentagon, which was built almost at the same time. As early as the 1950s,

it became clear that the National Airport would soon be unable to handle the increasing passenger traffic. Since there was little room for expansion at this location, ground was broken in 1958 for the construction of Dulles International Airport 40 km west of the White House on the border of the still very rural counties Fairfax and Loudoun in western Virginia. Since 1960, the population of Fairfax County has more than quadrupled to 1.1 million in 2010, and that of Loudoun County has even increased by more than 13 times to 325,000. Due in no small part to many investments by the government, many high-tech companies have located here. Both Loudon County and Fairfax County have repeatedly been the county with the highest household incomes in the USA in recent years. In Chicago, adjacent to O'Hare Airport, which for several decades had the highest passenger traffic in the U.S., Rosemont has become a major location for conventions with a large number of hotels. A similar development was seen in the neighborhood of Dallas-Fort Worth International Airport in Texas, which opened in 1974. In a world where more than 320 million people fly to meetings and other work-related gatherings each year, it makes sense to build conference centers directly at airports (Kasarda and Lindsay 2011, pp. 40–56, 93–94, 107; ▶ www.census.gov). Button and Stough (2000, p. 254) studied 56 airport hubs and found that around 12,000 more people are employed in the high-tech sector in the vicinity of airports than in regions without an airport hub but otherwise the same characteristics.

2.3 Sprawl

Sprawl is a term with negative connotations and describes the increasing expansion of settlement areas and their unplanned urban *sprawl* (◻ Fig. 2.9). As new

◻ **Fig. 2.9** View from the Willis Tower in Chicago over the *sprawling* urban landscape

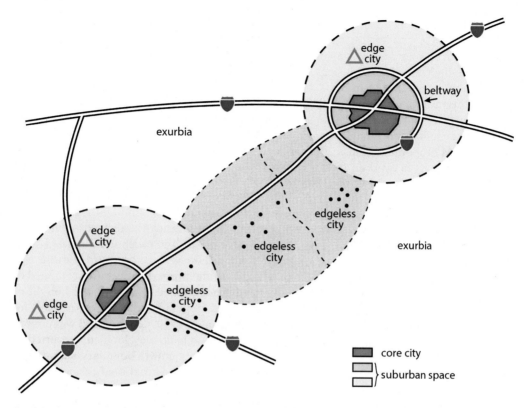

◘ Fig. 2.10 Spatial *sprawl*. (Adapted from Knox and McCarthy 2012, p. 110)

rings of owner-occupied homes are constantly being built on the periphery of suburban areas on ever larger plots of land, the population density of metropolitan areas decreases from the inside out. On the West and East Coasts, it is not uncommon for the spaces between individual metropolitan areas to be very densely populated, and *sprawl* has occurred, as illustrated by the BosWash, the band of cities from Boston to Washington, D.C. (◘ Fig. 2.10). Individual neighborhoods are connected by a network of expressways with up to 12 lanes, along which shopping centers and office parcs are located. Commuting distances are long, traffic and housing areas take up a lot of space, and environmental pollution is high. Even within housing estates, streets are exceptionally wide from a European perspective, although sidewalks are rare (Bruegmann 2005, pp. 17–20). Cities, whose development into large cities or cities with millions of inhabitants only began after World War II, were designed to be automobile-friendly across the board and are characterised by a particularly low population density. In the USA, more money is spent on building new roads than on maintaining old ones. Investors are thus encouraged to build new neighborhoods, shopping centers, and offices on the outer fringes of densely populated areas (Katz and Bradley 1999). Electricity costs are sometimes lower in suburban areas than in compact core cities. In New York City, they are driven up by high city and state taxes.

While this fills the public coffers, it does not lower per capita electricity consumption, which is already very low in New York City. At the same time, residents of huge houses in very remote new subdivisions pay little for their electricity because the cost is passed on to all users of a provider (Owen 2009, pp. 257–258). When one considers that land and real estate are cheaper and schools are better in suburban areas than in core cities, it becomes understandable why so many Americans live in the outskirts of cities. Ultimately, however, every taxpayer, as well as individual citizens, must pay a high price for *sprawl*. Building new roads is expensive, and commutes from home to work, shopping, or sports are becoming longer, more expensive, and more time-consuming (Gratz and Mintz 1998, p. 140). In many cities in the southwestern United States, water supply is a major challenge and further encourages horizontal urban *sprawl*. In Phoenix, a *Public Water Code* has regulated water rights since 1919, without which no new residential land may be developed. Because settlement in the region was very slow until the mid-twentieth century, water rights were predominantly granted to farmers. In recent decades, residential and commercial development has been preferred on former agricultural land with guaranteed water rights, rather than on still pristine desert land. In the Phoenix MA, this has created a checkerboard pattern of residential areas and desert areas (Keys et al. 2007, p. 143).

2

2.3.1 Countermeasures

Whether and how to curb urban *sprawl* is a controversial issue in the U.S. Unfortunately, as has been shown time and again, even measures that seem reasonable at first glance can have disadvantages. Even the expansion of public transport, which is supposed to save space for stationary and moving traffic, can promote *sprawl*. In 2000, the New York Metropolitan Transit Authority added two new stations to its network in the far north and east, about 130 km from Grand Central Station. Since the good connection enabled a quick commute to jobs in Manhattan, investors immediately designated building land near the new stations (Owen 2009, p. 136).

The many inner-city brownfield sites pose a particular problem. Residential and commercial areas are not systematically developed from the centre towards the periphery, but repeatedly skip over larger plots of land that are not developed. This is referred to as *leapfrog development*, because a frog also skips over larger areas, only to land on a point seemingly at random. Other brownfields have been created by deindustrialization. Because of the large supply of land, redevelopment of land rarely takes place, which is difficult anyway when the owner of the property has declared bankruptcy. Brownfields are treated like unwanted waste that blights the landscape, which is reflected in the term *drosscape* (*dross* = waste, *landscape* = landscape) for these areas. From an American perspective, the creation of *drosscapes* is considered inevitable as part of the development process of an urban landscape. In recent years, however, there has been a shift in thinking. It has been recognized that brownfields do not have to be permanent, but can be put to new uses. Between 1990 and 2005, more than 600,000 contaminated industrial sites in the United States fell into disuse, offering a variety of opportunities for future development. The federal government supports the revitalization of brownfields through a number of incentive programs, such as the conversion of land previously used for military purposes. In Denver and communities to the west in the foothills of the Rocky Mountains, some 215 km^2 that were heavily contaminated have been put to new uses. A paradigm shift has begun among citizens and local politicians. In the face of rising populations and major settlement pressures, even toxically contaminated brownfields are increasingly being put to new uses, with a positive impact on tax revenues for cities (Berger 2007). Nevertheless, investors often prefer building on newly developed land to the less calculable conversion of brownfields (Katz and Bradley 1999). The emergence of *edgeless cities* can also contribute to settlement densification, as new office buildings are often built on land that was previously *leapfrogged* as part of *leapfrog development* (Spivak 2008, p. 131).

Urban planning tries to limit urban *sprawl* with *smart growth* and *growth management*. A wide range of measures, such as redensification of existing residential areas, designation of small lots for residential development, creation of development corridors along public transport routes, or preference for mixed-use developments, can promote *smart growth*. Oregon was the first state in the U.S. to adopt a *growth management plan in* 1973 to limit uncontrolled land growth. In the meantime, many other states have followed Oregon's example in combating urban *sprawl*. The most important instrument is the introduction of *urban growth boundaries* (UGBs), which define the outermost limit for the expansion of the settlement body. Outside the defined boundaries, settlement is only possible in exceptional cases. However, growth boundaries have been criticized because of fears of rising land and property prices if unrestricted development land is not allowed to be designated on the periphery. In Portland, population density was actually lower in 2010 than it was in 1950; i.e., the UGB was not really successful. Although no development occurred immediately adjacent to the growth boundary, suburbanization could not be stopped as it shifted to more distant locations and to Washington State, which borders to the north (Cox 2011a; Cullingworth and Caves 2008, pp. 166–180; Ruesga 2000; The City Club of Portland 1999).

2.3.2 New Urbanism

New urbanism is an instrument of urban and regional planning and, in the broadest sense, includes all measures to reduce *urban sprawl* and the construction of neotraditional neighborhoods in order to increase the quality of life and the cohesion of the residents. In addition, sustainability is strived for. The ideas of *new urbanism* have been developed since the 1980s and institutionalized in 1993 with the founding of the non-profit *Congress of the New Urbanism*. According to the *Charter of New Urbanism*, the implementation of the far-reaching goals can take place at the micro, meso and macro level. Regionally, open spaces can be built on or subsequent densification of already built-up areas can be carried out. Small-scale structures should be preserved where they exist, and topographic features such as riparian areas or prominent ridgelines should be accessible to all citizens. In addition, parks that are close to residents should be created. At the intermediate level of a district or neighborhood, the goals are pedestrian friendliness, a mix of uses, a more heterogeneous composition of the population, and dense development. At the micro level, blocks, streets or individual buildings can increase residents' identification with their immediate living environment. Even in its initial phase, *new urbanism* was crit-

⬛ Fig. 2.11 Small-scale multifunctional building not far from the center of Austin

⬛ Fig. 2.12 Kentlands in the suburban area of Washington, DC

icized as being unaffordable, elitist or *boutique sprawl*. Information on the number of *new urbanism* projects implemented varies, but probably around 1000 measures have been implemented to date at all scales (Flint 2012; Schemionek 2005, pp. 68–70; ▶ www.cnu.com).

In the cities, more and more small-scale buildings with multifunctional use have been built in recent times in line with *new urbanism* (⬛ Fig. 2.11). At the same time, entire neighbourhoods or even larger settlements have been developed and built in suburban areas with *new urbanism* in mind. Small plots of land, a mixture of detached single-family houses, terraced houses and shops for daily needs as well as connections to public transport are important characteristics of *new urbanism* neighbourhoods. Shops, parks and public facilities should be within walking distance for all residents. Social contacts can be promoted by dispensing with swimming pools on private properties and having a large swimming pool for all residents. In some cases, private mailboxes are even dispensed with so that residents must regularly visit a central mailroom, where they can, at least in theory, get to know their neighbors. Streets and houses in the *new urbanism* developments are designed in the style of older neighborhoods. Narrow and winding streets prevent people from driving too fast. Trees provide shade, and garages are accessed via *back alleys* that run parallel to the streets between two rows of houses. Garbage cans and power cables, which in the USA are laid above ground on high poles, are also located here. As in the old days, many of the single-family homes of the *new urbanism* houses open onto the streets with small porches. It is hoped that residents will spend their evenings here, as in the old days, observing the hustle and bustle of the streets and socializing (Schemionek 2005, pp. 71–76; author's observations). Seaside in Florida, built in the 1980s by architects Andres Duany and Elizabeth Plater-

Zyberk, is considered the oldest *new urbanism* settlement, although it is not inhabited year-round, but only during the summer months. Celebration, also built in Florida by the Walt Disney Group in the 1990s, is probably the best-known development in this style. Probably the largest *new urbanism* development has been under construction since 2002 on the 19 km² site of the former Stapelton Airport in Denver. Within 15–20 years, 12,000 homes, office and retail buildings, and 4.5 km² of green space are to be built and landscaped here (Berger 2007, p. 70).

Kentlands, just outside Washington, D.C., is one of the best-known *new urbanism* developments in the United States (⬛ Fig. 2.12). In 1988, developer Alfandra purchased 142 acres of agricultural land in Fairfax County for the construction of Kentlands and commissioned Duany and Plater-Zyberk to create a new *urbanism-style* development for approximately 5000 people in 1600 housing units. The first model home opened in 1990, demand was high and Kentlands was completed about 10 years later. Kentlands consists of 12 separate neighborhoods that blend well into the hilly surroundings with mature trees and several ponds. The homes sit on relatively small lots, and the narrow and winding streets are indeed reminiscent of a traditional small town. Kentlands stands out pleasantly from other master planned subdivisions in the suburban area. However, not all of the goals of the *Charter of New Urbanism* have been realized to full satisfaction. Many things, such as the colorful shutters nailed to the facades next to the windows, which cannot be closed, are not convincing. The streets are mostly deserted, and Kentlands looks extinct. Porches facing the streets are not used. Also the desired heterogeneity of the population could not be achieved, because almost all residents are white and belong to an upper income class. Washington, D.C., which is about 25 km away, can be reached by public transport, but the trip takes about 90 minutes due to multiple changes. Therefore, the residents of Kentlands do not do without a car. A complete failure is the shopping street with small shops, whose success was threatened from the beginning. The developer had planned to build a closed shopping centre in Kentlands in order to secure his project financially with the income from renting out the shops. After a falling out with the investor of the planned shopping center, the discounter Wal-Mart wanted to open a hypermarket on the space designated for retail. Since Wal-Mart's strategy was not compatible with the goals of the *new urbanism*, the settlement of the low-cost retailer was fought by the residents. Eventually, they built an open shopping center with a variety of short- and medium-term vendors near the main shopping street. As is common in the USA, Kentlands residents drive to the shopping center, and the shopping street with rather unattractive shops is deserted even at peak shopping time on Saturdays (author's observation; ▶ www.kentlandsusa.com).

Other *new urbanism* neighborhoods are also convincing only at first glance. Residents rarely, if ever, use public transport, the population is homogeneous, a small-town idyll has at best seemingly emerged, and the social cohesion of the population is hardly greater than in conventional settlements. Critics describe the backward-looking architecture as kitschy or *Disneyfied*; moreover, the neighborhoods are neither new nor urban and correspond at best to a *new suburbanism. New urbanism* has become a marketing tool, and houses here are often more expensive than in other neighborhoods, creating islands of wealth with a homogeneous population. The high real estate prices even help to exclude certain groups of the population. It should be acknowledged, however, that the comparatively dense development of *new urbanism* neighborhoods limits sprawl (Knox and McCarthy 2012, pp. 339–340; Schemionek 2005, pp. 281–293).

2.3.3 *Sprawl* Versus Anti-sprawl

With growing environmental awareness, the number of opponents of unlimited urban sprawl has grown in the USA, but by no means everyone views *sprawl* negatively. The *anti-sprawl* movement, with the Sierra Club as its most important representative, advocates compact settlement construction and the greatest possible preservation of the natural landscape, while *pro-sprawl* supporters want to strengthen and guarantee the property rights of all individuals. Everyone should be able to profitably put as much land as they want to a new use (Berger 2007, p. 40; ▶ www.principlesofafreesociety.com). When new

neighborhoods are planned on the periphery, disputes often arise between opponents and supporters. In order to prevent construction, environmental organisations sometimes buy the land under discussion with the financial support of the opponents. Since this is only possible to a limited extent, the opponents and the municipalities, which are interested in increasing tax revenues, often reach a compromise. The construction of the controversial neighborhood is approved, but in return they only allow a limited number of houses to be built or compensatory land must be designated elsewhere. In the end, all parties involved are satisfied with this solution. The investors make good money, the municipalities receive high property tax revenues, and the local population is happy about the gain in open space and the influx of high-income new residents. From an environmental perspective, however, this practice is not to be welcomed, because the low population density in the new neighborhood promotes further urban *sprawl* (Rudel et al. 2009). On the other hand, it is positive that in some cities building plots have recently become smaller again and the proportion of terraced and multi-family houses is increasing. In addition, there is a redensification of built-up areas through the use of gaps between buildings (Bruegmann 2005, p. 61). A trend reversal can also be seen in the suburban area of the fast-growing MA Washington, D.C. Here, almost only townhouses (row houses) are built, since building land is expensive. Moreover, one may surmise that families with two working people have little interest in gardening.

2.4 Growing Cities

Particularly in the western and southern United States, many cities have experienced tremendous population growth in recent decades (◻ Fig. 2.13). The data must be put into perspective, however, because some cities have increased their parishes until the recent past, while others have not had this opportunity. Until the 1870s, it was common for towns to move their parish boundaries further and further into the surrounding countryside as their population increased. As the number of surrounding communities grew, this became increasingly difficult. In 1874, Boston became the first town to fail to expand its parish area as desired. Brookline, by its own account the "*richest town in the world*" at the time (cited in Jackson 1985, p. 149), successfully refused to be incorporated into Boston. Brookline's example was soon followed by other communities surrounding major industrial cities. Since the mid-twentieth century, cities in the old industrialized Northeast, as well as Los Angeles, have been able to expand their parishes at best moderately, while the areas of many sunbelt cities have grown many times

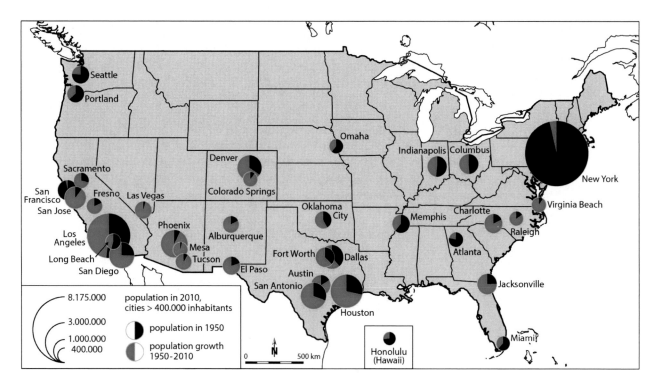

Fig. 2.13 Growing cities. (Data basis ▶ www.census.gov)

over. Phoenix has increased its land area by 150 times since 1950, and San Jose by 300 times. The cities with large land area growth have much lower population densities than most of the other cities. However, density also depends on location and historical development (**Table 2.1**). Overall, it is surprising that the population density of many U.S. cities is nowhere near as low as is often assumed. As expected, New York has by far the highest population density, but Boston, Chicago and Philadelphia also have a higher density than Munich, the city with the highest population density in Germany.

Cities whose population is growing rapidly have a positive net migration balance, as far more people move in than move out. Since the well-educated are more mobile than the poorly educated, they are particularly likely to relocate to cities with a large supply of jobs. This is often referred to as a brain gain. Austin, Atlanta, Boston, Denver, Minneapolis, San Diego, San Francisco, Seattle and Washington, D.C., have attracted particularly large numbers of highly qualified people. However, not all growing cities benefited equally from in-migration. Las Vegas attracts more high school dropouts than college graduates, and Phoenix is home to many construction workers and senior service providers (Harden 2004, pp. 178–179).

There are a number of different explanations for the significant increase in the number of inhabitants. Many new jobs have been created in the west and south of the country, where union influence and wages are low. There has been a large supply of cheap building land and few

constraints on attracting new businesses, and the military and aerospace industries have invested heavily in selected locations. Soft location factors such as sunshine and beaches and the absence of the long, harsh winters of the Northeast and Midwest added to the attractiveness of sunbelt cities (Conzen 1983, pp. 143–145). In Miami, the influx of Cubans was significant, while San Jose owes its rise to the development of Silicon Valley as the world's most important high-tech location. Houston has benefited from the development of vast oil fields in the Gulf of Mexico and has become a major center for global oil and petrochemicals. The population has grown from 0.6 million in 1950 to 2.1 million in 2010. Today, many major petroleum companies such as ConocoPhilipps, Halliburton and Marathon Oil have their headquarters in the Texas metropolis, complemented by research and production facilities. Oil in the Gulf of Mexico is important to the region, but other factors are also having a positive impact on investment. Low taxes and a widespread lack of union influence provide the basis for low wages and production costs. In addition, the U.S. has located research-intensive defense industries in the region and invested in infrastructure. Since land use plans are still lacking today, there are hardly any restrictions on the construction of new industrial plants or housing developments. However, the disadvantage is that a completely *sprawling* urban landscape has been created, and there is enormous dependence on the automobile (Hill and Feagin 1987; ▶ www.houston.com). San Diego has also ben-

2

▫ Table 2.1   District areas and population density of selected cities 1910–2010

City	Area in km²			Inhabitants per km²
	1910	**1950**	**2010**	**2010**
Baltimore	78	205	210	2597
Boston	101	119	124	5492
Chicago	479	576	590	4569
Dallas	41	290	881	1410
Detroit	106	352	350	2037
Houston	41	414	1554	1349
Los Angeles	220	1168	1214	3123
New York City	774	774	785	10,414
Philadelphia	337	331	347	4396
Phoenix	n.s.	44	1339	1080
Pittsburgh	104	135	142	2148
San Antonio	93	181	1194	1111
San Diego	192	256	418	1548
San Francisco	109	117	122	6598
San Jose	n.s.	44	458	2080
Seattle	145	184	218	2789
St. Louis	158	158	161	1981
Washington, D.C.	158	158	158	3804
For comparison:				
Hamburg			755	2347
Munich			310	4275

Sources: Jackson (1985, pp. 139, 141); ▶ www.census.gov

efited from investment by the military. Until the 1930s, the city was a tranquil coastal resort for wealthy tourists and senior citizens who enjoyed spending the winter here. In 1940, only 200,000 people lived in the city. After the Japanese invasion of Pearl Harbor (Hawaii) in December 1941, San Diego was developed into a major military base with the headquarters of the Pacific Fleet. Shipyards for military aircraft and warships were also built here. By 1950, the population had risen to 334,000. When some of the military was withdrawn at the end of the war, more than a few had hoped that San Diego could once again become a relaxing coastal resort. But by the early 1950s, the Korean War caused military importance and population numbers to continue to rise (Corso 1983, pp. 328–333). Today, San Diego is California's second largest city after Los Angeles, with a population of 1.3 million. The military is still a major employer, but biotechnology and telecommunications have also been able to gain a foothold here. In addition, the mild climate,

the fantastic location on the Pacific Ocean and facilities such as SeaWorld and the San Diego Zoo attract many tourists (▫ Fig. 2.14).

2.4.1   Challenges of Growing Cities

The influx of many people in just a short time presents cities with many challenges. New roads have to be laid, residential areas developed and schools built. At the same time, a large number of new citizens, who often come from abroad, have to be integrated. Rapidly growing cities also have disadvantages for residents. In Phoenix, Las Vegas, Atlanta or Dallas, it is almost unbearably hot in the summer, the streets are always congested, and the cities have no culture from the point of view of the residents of the Northeast. The question arises as to why so many people move to these cities. Glaeser (2011a, p. 183) doubts that the large supply of jobs alone is the decisive

Fig. 2.14 San Diego

Table 2.2 Prices for single-family houses in selected cities in 2012 (US dollars)

City/Region	Core city	Metropolitan area
Boston	359,000	314,000
Chicago	161,000	161,000
Detroit	29,000	79,000
Houston	160,000	175,000
New York City	454,000	342,000
Las Vegas	131,900	139,000
Los Angeles	425,000	405,000
Phoenix	121,000	153,700
San Jose	513,000	606,100

Source: Zillow Real Estate (2012)

factor. He sees low real estate prices and a low cost of living as an important reason for the attractiveness of many cities, which, however, react very differently to a large demand for housing. In Dallas, Houston, Phoenix and Chicago and many other cities, housing supply increases with demand because building permits are granted very generously. As the population grows, more homes are built and home prices remain stable. In New York, Los Angeles and Boston, increasing demand does not lead to increased supply, as hardly any new building permits are issued. Residential real estate is very expensive here (▢ Table 2.2). Prices are particularly high in Silicon Valley, where only 16,000 new single-family homes were approved for construction from 2001 to 2008. If 200,000 new homes had been approved, the cost of a home would be 40% lower (Glaeser 2011a, pp. 183–189;

Postrel 2007b). Gyourko et al. (2006) refer to cities that only the wealthy can afford to live in as *superstar cities*. These cities are characterized by an interesting mix of attractive leisure and productive work environments, as well as high incomes of residents. The high prices mean that only the wealthy can move to San Jose, Cupertino, Palo Alto or Mountain View, while poorer people concentrate on less attractive communities on the outskirts of Silicon Valley (▶ www.census.gov).

2.5 Shrinking Cities

Shrinking cities are characterized by high population losses in recent decades and are mainly found in the old industrialized northeast of the USA (▢ Fig. 2.15). In most cases, population decline is associated with a loss of jobs and a deterioration of the built environment. People have moved their residences to suburban areas or other regions for many reasons. They have left because jobs have moved to suburban areas or other regions of the U.S., or because winters are warmer and the cost of living is lower in the South and West. People have also moved away because the government has subsidized home construction in suburban areas. In addition, there was white fear of blacks and rising crime in the cities. The race riots of the 1950s and 1960s turned entire neighborhoods into war zones to be avoided. Residents of the core cities felt increasingly unsafe, or at least uncomfortable, in the increasingly unattractive streets lined with abandoned houses and lots. Graffiti and trash took over the cities, and those who could moved to suburban areas. The exodus was selective. Since it was preferentially the affluent, who were usually also the better educated and highly skilled, who moved to suburban areas or other regions, many shrinking cities suffer from a brain drain. The remaining population is often uneducated and poor, and often stays only because they cannot afford to move. Because of the oversupply of housing, properties can't sell despite low prices. In Detroit, there is no demand for single-family homes, even though the average purchase price in 2012 was only $29,000 (Zillow Real Estate). As population decreases, tax revenue decreases and less money is available for city infrastructure, social services, and schools, which are funded by local taxes in the United States. City lights must be kept on, streets must be cleaned, and emergency service plans must remain in place. Empty coffers force cities to lay off police officers despite high crime rates. Law and order can no longer be guaranteed. The cities get into a downward spiral that can hardly be stopped because no new citizens are moving in.

At the time of industrialisation, cities grew in rings from the inside outwards or along the main transport

2

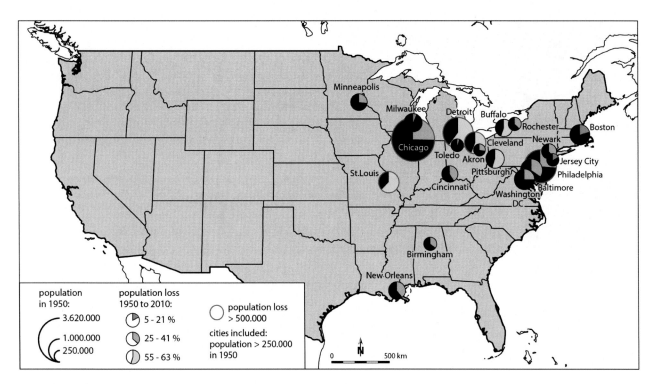

□ Fig. 2.15   Shrinking cities. (Data basis ▶ www.census.gov)

routes, and new districts were attached to existing ones. The depopulation of cities, on the other hand, takes place in a disorderly fashion. Inhabitants do not leave the outermost or central districts first, but preferentially abandon neighbourhoods with particularly large deficits. At the same time, the number of inhabitants in preferred neighbourhoods may still increase (Glaeser 2011a, p. 180; Harden 2004, p. 179; Kromer 2010, pp. 9, 78–79). In some cities, not only have the people left the cities, but even the dead or the buildings. Some of the families who left Detroit had their loved ones exhumed and buried in their new place of residence (Kasarda and Lindsay 2011, p. 185). In St. Louis, people demolished buildings that were no longer occupied in order to rebuild them elsewhere. In the 1980s, the "*Gateway to the West*" was the largest supplier of weathered brick, which was transported by rail or on the Mississippi River to the southern United States, where it was used to restore and build homes that would appear old (Jackson 1985, p. 218).

Many shrinking cities depended on a single industry, others even on a single company. Detroit was dominated by the automobile industry and Pittsburgh by steel production. In Flint, which is north of Detroit, about 80,000 employees worked for General Motors in 1968; by 2011, that number was down to 6000. At the same time, the population nearly halved from about 200,000 to just over 100,000 in 2010. In Rochester, NY, at the end of the nineteenth century, George Eastman had founded Kodak, a company that was extremely suc-

cessful for the next 100 years. When the city's largest employer by far announced the end of slide film production in early 2012 and filed for bankruptcy, fear reigned in Rochester as a population exodus was feared (The Economist 1/14/2012b). In many declining cities, the process of decay has now reached the point where the old downtown is no longer recognizable. In some former steel locations in the state of Indiana south of Chicago, the remnants of the buildings can only be recognized with a great deal of imagination as the former City Hall or Grand Hotel (observation by the author). However, shrinking cities offer artists plenty of room for design opportunities. In Detroit, a series of artworks have been created along Heidelberg Street primarily from materials that no one actually pays attention to. The so-called Heidelberg Project has since become a visitor magnet (Arens 2004) (□ Fig. 2.16).

2.5.1   Countermeasures

In the USA, shrinking cities do not receive any support from the federal government that is remotely comparable to the *Stadtumbau Ost* or *Stadtumbau West* programs known from Germany and that could help to cushion the negative effects of shrinkage. The cities are largely left to their own devices, as little help can be expected from the states or counties. Assistance in dealing with the problems of shrinking cities is available only from non-profit think tanks such as the Washington, D.C. based

◘ Fig. 2.16 The most famous house on Detroit's Heidelberg Street

Brookings Institute or university institutes such as the Metropolitan Institute at Virginia Tech University or the Shrinking Cities Institute affiliated with Kent State University in Cleveland, OH. These institutions conduct research in shrinking cities and develop recommendations for dealing with the cities' many shortcomings, but are unable to implement them themselves due to lack of funding and responsibilities. The American Assembly at Columbia University sees the poor image of the term *shrinking city* as a major problem and prefers the term *legacy city* for cities with declining populations. *Legacy* is meant to remind us of the glorious past, the historic building fabric and the cultural heritage of the cities from an American perspective. Moreover, *legacy cities* are characterized not only by weaknesses, but also by strengths such as the headquarters of major corporations, excellent museums, good hospitals and universities, and non-profit organizations. Therefore, in the view of the American Assembly, the conditions for positive development are not so bad (The American Assembly 2011, p. 5). Indeed, the Detroit Institute of Arts is one of the best museums in the U.S., as for about 100 years wealthy Detroiters have donated world-renowned exhibits. Much the same can be said for the museums of the city of St. Louis. For decades, attempts were made to breathe new life into shrinking cities. All measures were aimed at attracting people or new industries. Only recently have cities increasingly accepted that the number of inhabitants is unlikely to increase significantly in the short or medium term. New concepts are being developed to deal with the shrinking population and the resulting problems. Analogous to *smart growth*, there is talk of *smart shrinking* (Lunday 2009, p. 68).

Formerly a coal and steel hub, Youngstown, OH, whose population dropped by 100,000 from 1950 to 2010 to just 67,000, realized that the city would never return to its former glory and decided to demolish

vacant buildings in 2005. The new land use plan provides extensive open space for recreation and horticulture and is now considered one of the most innovative urban development plans in the United States (Gallagher 2010, pp. 15–16; Glaeser 2011a, p. 66). Very important is the cleanup of the city. Residents have become accustomed to the sight of abandoned houses and dilapidated properties over decades. The top priority, therefore, is to address these blights. Whereas previously federal funds were granted equally to rehabilitate dilapidated homes in all locations, today applicants receive money only if they are willing to move to selected "*healthier*" neighborhoods. In addition, investors who want to build new buildings in the city must focus their activities on certain neighborhoods specified in the plan. In the long term, the above measures should contribute to the creation of large contiguous open spaces and enclosed built-up areas where urban infrastructure can be concentrated. Youngstown wants to become a model for other shrinking mid-sized cities and polish up its poor image. Unlike in the 1950s, when entire neighborhoods were razed and residents forcibly relocated as part of the first urban renewal efforts, participation in the program is voluntary and well received by those affected (Gallagher 2010, pp. 17–18, 143; The American Assembly 2011, p. 8).

2.5.2 Vacancies, Derelict Land and Land Banks

Derelict land and empty houses are among the biggest problems of shrinking cities. In the face of declining populations and reduced economic activity, properties are first abandoned sporadically, quickly appearing unkempt and not infrequently used for waste of all kinds. Soon the houses, often built with wood, show signs of decay. The value of neighboring buildings, entire streets or even neighborhoods declines as the number of abandoned houses increases. Since vacant buildings attract criminals and gangs, cities board them up as quickly as possible. However, given the large number of abandoned buildings, cities are not keeping up. In Buffalo, NY, more than 10,000 vacant buildings blight the city and pose a safety hazard. In 2007, more than 40% of all shootings occurred here (Center for Community Progress 2012, p. 4). When property owners stop paying property taxes, the houses or lots sooner or later become municipal property. Within their means, municipalities demolish the dilapidated houses because they are unsightly and pose a hazard. The city of Detroit owns about 40,000 lots and houses that previously belonged to private owners. Given this large number, the indigent city is overwhelmed with maintaining or demolishing the homes. In New Orleans, with the assistance of the

2

Center for Community Progress, more than 28,000 lots and dilapidated homes had been assessed by the end of 2011. 2280 homes were demolished and just over 1000 homes that owners no longer cared about were seized and resold by the city. In 2011 alone, 1750 properties in New Orleans were cleared out, some with the cooperation of the owners. In Philadelphia, the Pennsylvania Horticultural Society cleared weeds and trash from thousands of abandoned lots, applied new soil, and planted sod and a few trees or shrubs. The derelict lots were surrounded with a simple wooden fence. The properties give a well-maintained impression, are no longer littered and do not negatively affect the value of neighbouring houses. In inner cities, larger brownfield sites are often used as parking lots, which are not particularly nice to look at, but are preferable to unkempt brownfield sites. The parking lots guarantee revenue that can be used to pay off property taxes. If one day an investor wants to build a high-rise at the site, the owner of the lot can become a rich man in one fell swoop (Center for Community Progress 2012, p. 3; Gallagher 2010, pp. 97–99, 136).

Some states have enacted laws or guidelines for the establishment of so-called *land banks*, which have the task of returning urban brownfields to productive use. In developing ideas for dealing with brownfields, they are supported by the Washington, D.C. based NGO *Center for Community Progress*. The first *land bank* was established in St. Louis in 1971, followed by Cleveland in 1976, Louisville in 1989, and Atlanta in 1991. Unfortunately, in cities with large population losses, demand for the brownfields is very low. Land is usually sold through auctions based on highest bids. Speculators often take over the land at a very low price. If they fail to sell the land at a higher price, they pay no taxes and the land reverts back to the cities' ownership after a short period of time (Gallagher 2010, pp. 136–140). Michigan-based Genesee County, which includes the shrinking city of Flint, took an innovative approach in 2002. Flint's brownfields are not transferred to city ownership, but to county ownership. Genesee County is also taking over land that has been abandoned by its owners in the suburban area. The latter are usually more valuable than urban brownfields. The Genesee Land Bank does not auction the brownfields located in the suburban area, but has them sold by real estate agents at market prices. Profits from these sales are used to demolish dilapidated homes in Flint and clean up abandoned properties. In addition, the *land bank* financially supports the renovation of homes in Flint if they are located in still-functioning neighborhoods to prevent homes with greater signs of decay from lowering the value of neighboring homes (Gallagher 2010, p. 70). If a brownfield site is next to a house that is still occupied and in good repair, its owner can buy the neighboring property for a dollar. The plan seems to be working. In the first 5 years after the Genesee Land Bank was established, proceeds from suburban sales were used to demolish about 1000 dilapidated homes in Flint and renovate another 300. Meanwhile, following the Genesee County model, similarly operating *land banks* have been established in other states such as Ohio, Georgia, and New York (Gallagher 2010, pp. 136–138; Lunday 2009, p. 68; The Economist 10/22/2011).

Framework Conditions for Urban Development

Contents

3.1 Neoliberalism and *Urban Governance*

The concept of neoliberalism was developed in the 1940s, but has since been modified several times and is implemented differently in different countries. Originally, neoliberalism denounced the liberal economic order of the nineteenth century, in which market forces ruled freely, and sought a new more just order. The drive for deregulation and privatization intensified after the 1973 oil crisis, when, in the view of many British and Americans, the Labour Party in power in Britain and the Republican Party in the United States had utterly failed. In Britain in the late 1970s, Margaret Thatcher, then leader of the opposition, with the support of several conservative think tanks, took up the theoretical concept of neoliberalism and made it the programme of the new government in the election campaign (Peck and Tickell 2007). In 1979, Thatcher became British Prime Minister and Ronald Reagan, who espoused the same ideology, was sworn in as the 40th President of the United States in 1981. Thatcher and Reagan believed they could increase profits by reducing government and administration and became the main proponent of neoliberalism, also known as the Washington Consensus. Since in the USA the desire for the greatest possible freedom of the individual and the associated fear of big government are deeply rooted in the country's history, the idea of neoliberalism, which sees the market rather than the state as the most important regulator of the economy and urban development, fell on fertile ground. The state does not govern directly, but only indirectly from a great distance. The state bureaucracy is cut in favour of private and semi-public decision-makers. The individual is expected to make decisions in his or her own favor, since it is assumed that what is beneficial to the individual benefits the community. Decisions are made on the basis of cost-benefit analysis rather than social considerations. Even the recipients of social benefits were convinced by this concept and the associated withdrawal from the welfare state, as they no longer wanted to be dictated to by the government and state social benefits were bad anyway (Leitner et al. 2007, pp. 4–7).

Based on the ideology of neoliberalism, since the early 1980s in the USA the federal government, individual states and local authorities have increasingly transferred public tasks to private companies or non-profit organisations, which today build roads, run schools, care for the homeless and socially disadvantaged, are responsible for local transport and even build and run prisons. Free market supporters support privatization because private companies can supposedly operate more cost-effectively and efficiently (Gotham 2012, p. 633). The impact on U.S. cities, many of which were in a downward spiral and lacked the financial resources to make important investments, was severe. Local politicians gratefully embraced the ideas of neoliberalism, as privatization and deregulation allowed for a retreat from municipal responsibilities and saved public coffers. At the same time, it was recognized that cities, like free enterprises, were competing for investment, economic success, and the brightest minds in other cities. New forms of *urban government* were developed, based on cooperation with the private sector in the form of *public-private partnerships* and *business improvement districts* with minimal interference by the municipality and maximum cost recovery by private investors. The American state, as well as the cities, withdrew from public housing and other assistance to the needy, but at the same time promoted the conversion of downtowns into storefronts for the region to attract private capital and new urbanites. The chapter is no longer used to address the concerns of all citizens equally, but favors the already better off. While in the liberal city every citizen still enjoyed equal rights, the neoliberal city deliberately excludes certain groups of the population. Permanent surveillance of public spaces and zero-tolerance policies allow for the expulsion of undesirables that might affect the overall impression (Brenner and Theodore 2002, pp. 20–25; Knox and McCarthy 2012, pp. 245–246; Leitner et al. 2007, p. 4).

As was most recently demonstrated in the 2011 presidential campaign, Republicans in particular are committed to reducing the tasks of the administration and government. However, this attitude is by no means shared by all Americans. Opponents of privatization fear that private companies wield too much power, operate for profit even in the event of disasters, incur high costs in the long term, and have too little commitment to the common good. In addition, there are repeated reports of corruption and illicit advantage-taking by private companies (Gotham 2012, p. 633). The Democratic President Barack Obama actually pursues a different ideology, but is repeatedly slowed down in the implementation of his program by the opposition, since the Republicans have the majority in the House of Representatives of Congress and block many bills.

3.2 Mayors

Deindustrialisation, deregulation, globalisation, neoliberalism and suburbanisation have confronted cities with major tasks, which mayors are responsible for managing (◘ Fig. 3.1). Cities today are to be managed like companies, of which mayors are the managers. Their vision and drive, but also their failures, have had a major impact on the development of many cities. Michael Bloomberg, mayor of New York from 2002 to 2013, ran

☐ Fig. 3.1 New York Mayor Bloomberg wanted to address all groups of the population

the city pragmatically like the company he founded in the early 1980s, Bloomberg News, which made him a billionaire many times over. Because of his wealth, he was considered incorruptible, and he also submitted to few constraints in partisan politics. Long a Democrat, Bloomberg switched to the Republicans in 2001 to boost his chances in the election for the city's highest office; but declared himself an independent in 2007. He wanted to solve problems as effectively as possible and advocated unpopular measures single-handedly if necessary. When necessary, he raised taxes or banned oversized sugary drinks; at the same time, he put in bike lanes and lobbied against uncontrolled gun ownership. However, Bloomberg must be accused that his activities were not always democratically legitimate (Barber 2013, pp. 25–28). After three terms in office, he was unable to run for re-election, which was won overwhelmingly by Bill de Blasio in the fall of 2013. DeBlasio, a Democrat, announced in the election campaign a new style of politics, the elimination of social injustice and the fight against the increasing income disparities and the housing problem in New York. Whether these goals will be realized remains to be seen (New York Times 2013 05/11/13).

Clark et al. (2002, p. 52 f.) studied the policies and assertiveness of mayors of selected large cities in the 1990s. The cities that developed particularly successfully were those that placed a great deal of emphasis on improving the school system and on cleanliness and that expanded cultural and recreational offerings. Basically, there is no mayor who works to the satisfaction of all citizens as well as the press and opposition, and mistakes are almost impossible to rule out given the responsible office and the wide-ranging demands. Indeed, for decades the spectrum of mayors has been wide, ranging from corrupt representatives who have criminally lined

their own pockets, to the hardliner Rudolph Giuliani in New York, and the comparatively successful Richard M. Daley, who guided the fate of the city of Chicago from 1989 to 2011. It has even happened that mayors have been arrested and convicted while in office. Marion Barry served as mayor of Washington, D.C., beginning in 1979. After being sentenced to prison for drug abuse, he was forced to resign the office in 1991, but held it again from 1995 to 1999. In addition, other local politicians were or still are often corrupt, as are the police or even the police chief. In Chicago, from 1971 to the late 1990s, 26 aldermen were sentenced to prison for bribery or similar offenses (Clark et al. 2002, p. 507).

Since the 1990s, a new generation of mayors has emerged in some cities, ready to take responsibility for themselves. Stephen Goldsmith, mayor of Indianapolis from 1992 to 2000, and Bret Schundler, mayor of Jersey City from 1992 to 2001, have succeeded in reducing crime rates, improving the school system, providing jobs for welfare recipients, making public services more effective, and attracting private investment (Malanga 2010, p. 1). Newark, which had been dominated by corrupt politicians for decades, has also finally seen a turnaround. Founded by Puritans as early as the seventeenth century, the city had developed into a major industrial city since the mid-nineteenth century, benefiting from good access to the rail network and proximity to New York. As industrialization increased, powerful labor unions had emerged in Newark that not only fought for workers' rights and wages but also exerted great influence on city politics. At the time of Prohibition (1920–1933), nationally known bootleggers plied their trade in Newark, and corruption continued to be rampant in the decades that followed. By 1950, when the city's population was 440,000, city expenditures were double those of other cities due to corruption in Newark. Understandably, beginning in the late 1950s, many businesses relocated to southern states and foreign countries where production could take place under better conditions and with lower labor costs. At the same time, more poor blacks from the rural areas of the Southern states flocked to Newark in search of work. The mostly uneducated blacks were competing with the displaced domestic labor force for a declining number of jobs. Mayor Addonizio (1962–1970) was on the take and had effectively abandoned the city to organized crime. Much the same was true of two of his successors and the police chief. Change did not occur until Cory Booker became mayor of Newark in 2006. He cut jobs in the bloated administration and, among other things, cut in half the wage subsidies for the overgrown and corrupt city housing authority. Mayor Booker saw it as one of his most important tasks to reform the ailing school system. His plans were so compelling that Facebook founder Mark Zuckerberg

3

donated $100 million to the cause after a visit to Newark (Malanga 2010, pp. 1–7). Meanwhile, Newark is thriving after decades of downturn, once again benefiting from nearby New York. In Detroit, Dave Bing, mayor since 2009, has taken similarly drastic measures. Like Booker, he believes that to manage a city effectively and competently is to establish trust in politics and administration. In Detroit, however, the problems were so great that failure was virtually inevitable.

3.3 Globalization

Today, cities are more than ever embedded in the global economy and dependent on global restructuring processes. They form nodes of transport and communication in a global network that is subject to constant change. The future of a city depends on its role in the global network. Global interdependencies are, of course, not new, for even the Hanseatic or colonial cities were linked to other cities and competed with them. These cities used to be referred to as world cities, but since the 1980s the term *global cities* has become accepted. In the face of increasing globalization, made possible by the liberalization of trade in goods and services and better means of communication, an international division of labor has developed for both production and services. Research and development takes place in developed countries, while production takes place in low-wage countries. *Global cities* concentrate high-value services, are home to the headquarters of transnational corporations and international institutions, and are characterized by a rapid growth of the service sector. At the same time, they are important transport hubs and destinations for national and international immigrants. Global functions need not be limited to core cities, but can encompass larger regions known as *world city regions* (Gerhard 2004, pp. 4–6).

Globalization has not only fostered competition between cities, but has also created new inequalities within cities. In *global cities,* many jobs have been created for the highly skilled with almost astronomical salaries, while at the other end of the scale, the number of jobs for the low-skilled, who receive only the minimum wage, has increased even more. With deindustrialization, many people in the U.S. have lost a secure job with union-guaranteed rights (Sassen 1994, pp. 241–247). The numerous jobs for building cleaners or the *mcjobs* referred to in reference to the fast food chain McDonald's do not offer these securities. At the same time, managers in the FIRE sector (finance, insurance, real estate) draw gigantic bonuses of sometimes several million U.S. dollars in addition to their high salaries, which they like to invest in luxury cars and apartments at the end of the year.

Global city research in recent decades has addressed very different issues such as the influence of globalization on urban development, the specialization of metropolises, social problems or the importance of non-economic networks formed by non-governmental organizations. On the homepage of the *Globalization and World Cities* (*GaWC*) *Research Network* of the British Loughborough University alone, 428 research papers are currently published. Particularly frequent are analyses of the relationships between cities and their intensity, or attempts at hierarchisation. Representative of the many studies on these two issues are those by Beaverstock et al. (1999) on internationally active companies in 263 cities and by Castells (2004) on the network society and communication technology. Beaverstock et al. (1999) quantified the size of the auditing, finance, advertising and legal services sectors and concluded that the size of these services is particularly large in New York, London, Paris and Tokyo. They refer to these cities as *alpha-cities,* followed by *beta-* and *gamma-cities* at the lower levels of the hierarchy. In the industrial age, railways, ocean-going ships or telegraph cables connected the world. In the globalized world, digital communication networks have become more important. Networks are an efficient form of organization because they are flexible, adaptable and survivable. Networks have no center, only nodes. All nodes are important to the functioning of networks. If a node loses importance, part of its functions can be taken over by another node in the network. It is even possible for a node to drop out of the network and be replaced by another node. In the past, the flows of goods played a crucial role in measuring the relationships between world cities, whereas today the intensity of information flows, which are becoming ever faster and more extensive, is far more significant. The emergence of global financial centers, which are an important feature of *global cities* at the top level of the hierarchy, would not be possible without modern communication networks (Castells 2004, pp. 3–28). In English, a distinction is made between the terms *hub* and *node,* both of which are usually translated as node in German and used synonymously. In fact, however, *hub* is of higher value and represents a *hub* for the exchange of goods or information, while *node* only describes the place where goods or information meet (Schmidt 2005, p. 287).

3.4 Vulnerability

Natural disasters triggered by earthquakes, hurricanes, tornadoes or tsunamis are not uncommon in the USA (◘ Fig. 3.2). In addition, there are disasters triggered by people through carelessness or by violent terrorists.

Hurricanes and earthquakes have repeatedly caused more or less widespread damage even in large cities. On

Fig. 3.2 Warning of tsunamis on the Pacific coast

the West Coast, earthquakes are a latent danger. The greatest damage was caused by the 1906 earthquake in San Francisco. Since then there have been a large number of smaller quakes in which houses or two-story streets collapsed or deaths and injuries were counted. While there are now many regulations for earthquake-resistant construction, comprehensive protection against this natural event is not possible. Although it is considered certain that a major earthquake will occur in the not-too-distant future due to the tectonic features of the region, the Pacific Coast continues to enjoy great popularity among residents and visitors. The cities on the East Coast and Gulf of Mexico are regularly hit by hurricanes in the fall, but they always announce themselves a few days in advance, even if the strength and exact location of the landfall are unpredictable. Strong winds and water repeatedly cause very extensive damage, while the number of casualties and deaths is usually low, as many coastal residents flee inland in the face of approaching storms or the authorities order the evacuation of entire areas of land. More recently, Hurricane Hugo (1989) in the two Carolinas, Hurricane Andrew (1992) in the southern tip of Florida, and Hurricane Katrina (2005) in New Orleans caused extensive damage. At the end of October 2012, Hurricane Sandy spread great terror in New Jersey and New York. When the storm hit the coastal region, it was no longer a hurricane, but "only" a tropical storm, but the damage was unusually extensive for several reasons, as large parts of New York City were submerged. Parts of Lower Manhattan, the Rockaways in Queens, Coney Island in Brooklyn, and Staten Island were still official disaster areas months later. Damage was particularly extensive in Atlantic City in New Jersey, where the center of the storm was located. Hurricane Sandy killed 43 people, injured many more, and caused billions of dollars in damage. The death toll would have been lower, however,

if some 375,000 people in New York had not resisted the evacuation ordered by Mayor Bloomberg. In view of global warming, it cannot be ruled out that the number of hurricanes will increase in the future (Gelinas 2013).

Tornadoes also develop very high wind speeds and, although they only last a few minutes, can cause very extensive damage on a small scale. They occur particularly frequently in *tornado alley* in the states of Oklahoma, Texas, Nebraska and Kansas, as well as in neighbouring states. Protection against the very fast-forming tornadoes is hardly possible. Since residential buildings in the USA are predominantly constructed of wood, tornadoes often destroy entire streets, city districts or small towns. On the coasts, there is also a latent danger from tsunamis, which can destroy entire swaths of land within minutes or become a major hazard. In addition to the above-mentioned natural events, landslides or fires often cause major damage. It is not uncommon for towns in densely forested regions to be threatened by fire when dry trees catch fire from lightning strikes or arson. Santa Ana winds pose a particular danger. These very warm downdrafts move from the Great Basin toward the Pacific Coast in late fall, often igniting the dry forests. Los Angeles, and San Diego in particular, are repeatedly exposed to the approaching conflagration. In addition, there have been repeated incidents of extensive damage, both man-made and arson.

When Chicago was largely destroyed by a fire caused by carelessness in 1871, and when San Francisco was hit by a devastating earthquake in 1906, there was no doubt whatsoever that both cities would be rebuilt, for they were in a period of economic boom at the time of destruction. In fact, most cities emerged from the crises remarkably quickly. In Chicago, more than 100,000 residents had been left homeless. By 1880, new housing had been created not only for those affected by the fire, but for an additional 200,000 new citizens. San Francisco was also rebuilt very quickly (Vigdor 2008, p. 136). However, after the destruction of New Orleans in 2005 by Hurricane Katrina, politicians and environmentalists seriously discussed foregoing rebuilding at the site. At the time, New Orleans was a very troubled city with a population that had been shrinking for decades, a lack of jobs, high poverty, and high crime rates (McDonald 2007; The American Assembly 2011, p. 3).

Globalization and technological progress have increased the vulnerability of cities, as the terrorist attacks of September 9, 2001, on the World Trade Center in New York and the Pentagon in Washington, D.C., demonstrated most painfully. Although skeptics have heralded the end of the skyscraper and predicted a new form of city (Davis 2001), the southern tip of Manhattan is more attractive than ever now that reconstruction is nearly complete. Today, however, danger to urban societies is also posed by pathogens that can travel

3

from continent to continent in a matter of days given global travel and world-spanning air links, as the rapid spread of the SARS viruses in 2003 demonstrated (Keil 2011, pp. 54–55). Fear and terror can also be spread by intentionally circulating pathogens, such as the anthrax bacteria sent through the mail in Washington, D.C., in late 2001. All cities with subways fear attacks on crowded trains or stations. Since New York has the largest subway network in the country and has a large global footprint, the city is especially concerned. Car tunnels and bridges into Manhattan and the drinking water supply are also major danger points. Water is supplied to the city through only three tunnels. Contamination with pathogens could threaten the health of the entire population, and closing the tunnels would make the city completely uninhabitable (Owen 2009, pp. 30–31).

Although there is no doubt that California is bracing for a major earthquake with perhaps hundreds of thousands of deaths and barely imaginable repercussions should the epicenter be in Los Angeles or San Francisco, and hurricanes will continue to wreak havoc on the East Coast, Florida, and the Gulf of Mexico in the future, cities in these locations are growing almost unchecked. Repeatedly, after major natural disasters as well as terrorist attacks, it has been shown that people roll up their sleeves and the economy recovers within a few years. Whether this will be the case in the future remains to be seen (Gong and Keenan 2012, p. 374; Owen 2009, pp. 30–31).

3.5 Urban Planning

The typical American, as well as local communities, reject or even fear any government interference in their affairs. Land is considered an economic asset like any other, and there is widespread consensus that an owner should be allowed to make the maximum possible profit from his land. This desire is respected by the government in Washington and by the states, which, with a few exceptions such as coastal and environmental protection, impose few requirements on municipal land use (*zoning*). Municipalities are largely free to decide for themselves whether and how to engage in residential development, building development, or redevelopment or revitalization of existing buildings or neighborhoods, and what form of citizen participation to allow (Cullingworth and Caves 2008, p. 17). Urban planning policies vary accordingly across major U.S. cities (**◻** Fig. 3.3). In some cities, such as New York and Los Angeles, urban planning and development have a long history. In Los Angeles, the *Residence District Ordinance* in 1908 was the first in the U.S. to mandate spatial separation of industrial and residential areas (Kotkin 2001, p. 3). In 1916, New York

◻ Fig. 3.3 Historic buildings with new construction in Washington, DC

became the first U.S. city to adopt a comprehensive land use plan, which is regularly revised and has since been expanded to include a large number of other requirements, such as the designation of *historic districts* (see below) (Cullingworth and Caves 2008, pp. 68–69). Houston, still the fourth largest city in the country, does not have a land use plan to date. While there are stipulations in Houston on the density of development allowed in selected neighborhoods or on the minimum number of parking spaces in the case of new construction, overall urban influence on land use in the Texas city is extremely limited. Between the two extremes of New York and Houston, there is a wide range of urban planning approaches.

The fragmentation of metropolitan areas and the large number of authorities involved in the planning process make orderly spatial development difficult. In addition, the individual municipalities are in competition with each other. This becomes clear when a shopping center is to be built in a region. Counties and municipalities fear an outflow of purchasing power and the loss of sales tax revenues if the center is located outside their own borders. Existing land use plans are readily overridden and further incentives are created to attract the shopping center. Since there is no state zoning ordinance whose requirements must be met, municipalities have a great deal of leeway. Private investors exert far more influence on urban development in the USA than in Europe. Many projects are negotiated between investors and the city and approved on a case-by-case basis. Citizen groups pursuing goals such as NIMBY (*not in my back yard*) or BANANA (*build absolutely nothing anywhere near anybody*) and politicians afraid of making decisions unpopular with voters (NIMTOO = *not in my term of office*) can delay the planning process by years or even decades (Cullingworth and Caves

2008, p. 80; Dear and Dahmann 2011, p. 75). In the San Fernando Valley, which is part of the Los Angeles MA, residents have protested very vigorously since the 1970s against the expansion of mass transit, the planned routes of which they felt disturbed them, even though this was the only way to avoid the threat of gridlock in the region in the long term (Davis 1992a, pp. 203–205).

In the nineteenth century, the focus was on redressing grievances, as large numbers of poor immigrants and industrialization had led to disastrous conditions and slumming in many places. The redress of grievances was limited to churches, businesses, or private individuals such as Jane Addams, who, along with Ellen Gates, opened Hull House on Chicago's Near West Side in 1889. The complex soon took up an entire city block. The 13 buildings provided housing for single working-class women and also included a gymnasium, library, boy's club, kindergarten, and more. Following the example of Hull House, nearly 500 similar institutions for the socially disadvantaged were established in the USA (Chicago Historical Society 2004, keyword: Hull House). Government support for housing did not begin until the USA implemented a series of economic and social reforms as a result of the recession of the 1930s. On the initiative of President Roosevelt (1933–1945), the *Federal Housing Act* was passed in 1937, providing $75 million in federal funding for the construction of housing for the working poor. Initially, the program was considered very successful, for by the end of 1938, 221 local authorities had been established throughout the country to promote public housing and some 130,000 units had been built in 300 different projects. However, especially in suburban areas, many communities resisted building housing with public funds, which they could not be forced to do. Moreover, for every dwelling constructed with public funds, one dwelling that was no longer habitable was to be demolished, thus contributing to the elimination of slums. As a result, the number of dwellings did not increase. The refusal of municipalities and the lack of poor housing stock in suburban areas led to a concentration of public housing in the core cities. More public housing apartments were built in Newark, NJ, than in any other city in the U.S., while adjacent municipalities vehemently resisted housing for the poor. By 1970, Newark was among the most troubled cities in the United States. Although the publicly built housing was initially intended to be temporary for more or less destitute residents who were willing to work, more and more people, who were also predominantly black or Hispanic minorities, were living permanently in these houses. The settlements were referred to as *the projects* and had a decidedly bad reputation. They became a gathering place for the unsuccessful and often criminals. Notorious was the Pruitt-Igoe housing development, completed in 1956 north of downtown St. Louis. There were nearly 3000 apartments in 33 eleven-story buildings. Inhabited almost exclusively by blacks, the apartments quickly became a gathering place for drug abuse and crime. The complex was demolished in the early 1970s. The largest public housing project in the USA was the Robert Taylor Homes in Chicago, completed in 1962 with many structural defects, comprising 28 high-rise buildings and 4415 apartments. As a result of deindustrialization, many of the tenants lost their jobs and any prospect of a new job. In addition, many apartments were overcrowded; at times as many as 27,000 people lived in the complex, increasingly living off informal and often illegal activities such as drug dealing. Repeated attempts by the Chicago Housing Authority to rid the housing complex of crime, such as the *Clean Sweep anti-crime campaign* of the early 1990s, were a complete failure. The Robert Taylor Homes were demolished between 2000 and 2005 (Chicago Historical Society 2004, keyword: Robert Taylor Homes). Since President Reagan drastically reduced funding for public housing in the 1980s, only very sporadic housing projects have been realized. With just under 180,000 apartments in 334 developments, the supply today is by far the greatest in New York. New York projects are likely to be home to far more than the 400,000 registered residents, as several families often illegally share an apartment due to the city's housing shortage. The New York City Housing Authority (NYCHA) is the largest public housing authority in the United States and the largest landlord in New York City (Hahn 1996a, pp. 118–119; Jackson 1985, pp. 219–228; ▶ www.nyc.gov).

3.5.1 Growing Influence of Private Investors

The *Housing Act* of 1949 had already initiated the move away from state subsidies. In future, housing construction was to be taken over preferentially by private investors. At the same time, the establishment of *public-private partnerships* between the federal government, municipalities and private investors was encouraged, which were to carry out the demolition of slums and problematic neighbourhoods and their redevelopment with high-yield and high-tax properties. The trend toward privatization of redevelopment projects was reinforced by the *urban-renewal* legislation of the 1950s. Private investors had wide latitude in defining and delineating redevelopment areas. In many cities, redeveloped areas were used to build sports stadiums, large multifunctional buildings such as Boston's Prudential Center with several hotels and a shopping center, or Government Center with City Hall and city offices. In other inner cities, problem neighborhoods were demolished and parking lots were created on the brownfields, which are still waiting for an investor today. Slums were eliminated without new housing being built for the homeless population. In many cases, luxury apartments were built on the brownfields for wealthy buyers and tenants. The *Model Cities Bill*, passed in 1966 under President Lyndon B. Johnson (1963–1969), took a broader approach than previous programs. Comprehensive rehabilitation of the built environment, economy, and social structures was to take place in selected model cities with technical and financial assistance from the federal government. The comprehensive program proved difficult to implement for many reasons. In 1974, it was replaced by the *Community Development Block Grant Program* (CDBG), managed and coordinated by the U.S. Department of Housing and Urban Development, which still exists today. The CDBG program provides a wide range of opportunities for urban revitalization and enhancement. These include anti-poverty measures, infrastructure development, or housing development. Ultimately, however, the main purpose of the program is to create incentives for private investment. Most of the funds flow into inner-city projects, while the socially vulnerable benefit little from CDGB (Cullingworth and Caves 2008, pp. 296–299; Schneider-Sliwa 2005, pp. 161–179). In February 2009, at the instigation of President Obama, the *American Recovery and Reinvestment Act* was passed to combat the American economic crisis (◘ Fig. 3.4). It supports a wide range of projects, such as the expansion of the education and healthcare system, tax relief for low-income earners, aid for farmers in the event of natural disasters, but also programmes to expand municipal infrastructure, revitalise neighbourhoods or build housing for the socially disadvantaged. By the end of 2012, $849 billion

had been disbursed. Although the focus is on stimulating the U.S. economy, the law gives local governments the opportunity to prioritize and implement measures that could not have been accomplished with their own funds. In the MA Honolulu on Hawaii, which suffers from constantly congested roads, the construction of a rail transit system is being funded, and in San Francisco and Boston, windows and roofs in social housing estates are being financed. The use of the funds is transparent and can be viewed in detail on a homepage of the U.S. government (▶ www.recovery.gov).

3.5.2 Urban Renewal Versus Urban Preservation

In the USA, there has been a dispute for decades as to whether unconditional urban renewal or the preservation of existing structures is more important. This is especially true for New York. In 1908, at the height of the railroad era, Pennsylvania Station was opened on 33rd St. The building, with Doric columns and a waiting hall in the style of the Baths of Caracalla in Rome, was no longer fit for purpose just half a century later. In the late 1950s, the Pennsylvania Railroad planned to replace the old station with a modern functional building. A 34-story office building was to be built above the station in order to generate a better return on investment. Despite fierce protests, the magnificent Beaux-Arts building was demolished in the mid-1960s (Glaeser 2011a, pp. 148–149). At the same time, under longtime city planner Robert Moses, New York was being transformed into a car-friendly city and old buildings were being demolished for new expressways, which also led to protests. Since the late 1920s, there had been talk of building a ten-lane expressway between the bridges that cross the East River and the Holland Tunnel, which runs under the Hudson River and connects Manhattan to New Jersey. After the passage of the *Federal Interstate Highway Act* in 1956, which provided for up to 90% federal funding for expressways, implementation of the plan became within reach. In 1959, the City of New York published a plan that called for the demolition of 416 buildings containing approximately 2000 residences and at least 800 commercial and industrial buildings. SoHo (South of Houston St.) and parts of the adjacent neighborhoods of Little Italy and Chinatown would be wiped off the face of the earth (Gratz and Mintz 1998, pp. 295–298). In the 1960s, New Yorker Jane Jacobs was probably the best-known opponent of Robert Moses's urban renewal. The two were adamantly opposed to each other. Jane Jacobs' book *The Death and Life of Great American Cities,* published in 1961, marked a turning point in the development of North American

◘ **Fig. 3.4** Work under the *American Recovery and Reinvestment Act* in Detroit

Fig. 3.5 Houses at Washington Square

cities, but remains highly controversial today (Husock 2009). The activist lived with her family on Hudson St. in New York's West Village not far from Washington Square (Fig. 3.5) in Manhattan and watched as well-functioning and vibrant streets and neighborhoods were demolished for the construction of large-scale buildings and new traffic lanes, losing their former diversity and vibrancy. She criticized the creation of monofunctional neighborhoods and fought, sometimes very successfully, the demolition or cutting up of old neighborhoods. Jane Jacobs organized the successful protest against the construction of the proposed ten-lane expressway (Glaeser 2011a, p. 145; Zukin 2010, pp. 13–14). Very early on, she recognized that a mix of different uses enhanced the vibrancy and creativity of the city. In small-scale neighborhoods like Greenwich Village, small shops were located on the ground floors, and residents knew each other. Neighbors and retailers jointly patrolled the street or sidewalks, contributing significantly to their safety. The use of sidewalks has received the greatest attention from Jacobs (1961, pp. 29–88). Sidewalks must be used by many people and be easily visible. Empty sidewalks between large monofunctional buildings that do not open to the street do not fulfill this function. Strangers meet in cities. Sidewalks are only safe if neighbors and merchants always keep an eye on them. The "*eye upon the street*" is the most famous metaphor from the work of Jane Jacobs (1961, p. 35): "*… there must be eyes upon the street, eyes belonging to those we might call the natural proprietors of the street*". Glaeser (2011a, p. 146), while criticising Jane Jacobs, concedes that this is not wrong. However, he believes that high-rise buildings can also create a safe environment. Since a lot of people work or live in high-rise buildings, the sidewalks are highly frequented and thus create a sense of safety for all users of the streetscape. In addition, there is nothing to prevent shops and restaurants from being located on the ground

floors of high-rise buildings as well. For this reason, the skyscraper-lined streets of Midtown Manhattan are also very safe. Glaeser (2011a, pp. 146–148) believes that Jane Jacobs did more harm than good to New Yorkers. The disadvantage, he argues, is that comparatively few people live in the low-rise houses in the neighborhoods Jacobs saved. It has contributed to a shortage of housing in Manhattan, the result of which is high rents. A city like New York, he said, depends on high-rises to provide affordable housing for all segments of the population. High-rise buildings, not small buildings with only a few apartments, are a guarantee for a vibrant city. Sharon Zukin (2010, p. 17) criticizes Jane Jacobs for similar reasons. The latter, she argues, wanted to conserve the city she knew and failed to recognize the influence of capital on urban development. Schwarz (2010) criticizes Jacobs for being one of the first generation *gentrifiers* and for contributing to the transformation of the West Village herself. Others praise the influence of Jane Jacobs on the development of the North American city, saying that she initiated a paradigm shift from a car-oriented to a livable city (Sorkin 2009, pp. 22–23). SoHo would not exist in its current form without Jane Jacobs if the proposed expressway had been built (Fig. 3.6). She had worked out for the first time what a living city should be like and never tired of proclaiming it. Jane Jacobs had recognised the potential of SoHo and successfully led the protest against conversion. The events that took place in SoHo in the late 1950s and early 1960s are now known as the "*SoHo syndrome*". Not only was it henceforth difficult to dissolve established small-scale structures in favor of large-scale projects in other locations in New York, but the "*saved*" neighborhoods also liked to give themselves names reminiscent of the SoHo model. The successful fight against the construction of the Lower Manhattan Expressway quickly had an impact on

Fig. 3.6 Broome Street was to be converted into an expressway in the 1960s

neighboring neighborhoods such as TriBeCa (Triangle below Canal St.), Noho (North of Houston St.), EsSo (East of SoHo), and the East Village, all of which were placed under change protection. Soon these measures were extended to Brooklyn neighborhoods such as Dumbo (Down under the Manhattan Bridge) and the historic working-class neighborhoods of Greenpoint, Northside, and Williamsburg, where residential and commercial uses had always existed in close proximity. Following the New York model, Denver's Lower Downtown became LoDo, Seattle's South of the Dome became SoDo, and San Francisco's South of Market became SoMa (Gratz and Mintz 1998, pp. 302–303).

In response to the protests of Jacobs and other activists, New York Mayor Robert Wagner founded the Landmarks Preservation Commission in 1962, which was predictably criticized by the real estate industry, which feared rigid building regulations. By the mid-1960s, some 700 buildings in New York were already protected by the commission, and by 2010 there were more than 25,000. To date, about 100 *historic districts* have been designated in the city (◘ Fig. 3.7). South of 96th St. in Manhattan, *historic districts* cover more than 15% of the built-up area. Any changes here must be approved by the Landmark Preservation Commission (► www.nyc.gov/landmarks). Between 1955 and 1964, the city approved the construction of 11,000 homes, but between 1980 and 1999, it approved only 3120 homes. With a rising population, housing in New York is hotly contested and rents have increased 284% in constant

prices from 1970 to 2000. New York's scarce land is not being used to its fullest potential. In a 40-story high-rise, a 110 m^2 apartment accounts for barely 3 m^2 of floor space, far less than in a protected four-story building where demolition is not allowed. Yet new apartments of this size in New York often cost more than $1 million. By contrast, in Houston, where residential construction is hardly limited, a new 230 m^2 building costs only $200,000. Only the wealthy can afford to live in New York's *historic districts*. Residents of the protected neighborhoods are more likely to have a college degree than the average New Yorker, and the proportion of whites is higher than in other neighborhoods. But even in neighborhoods that are not protected by the Landmarks Preservation Commission, prices are high because construction of new buildings in *historic districts* is limited. New York today recommends itself only as a place to live for the rich, who can pay any price for their homes, and for poor immigrants. The latter move into the unrenovated apartments in poor locations with access to public transport. They cannot live in the suburban area because they cannot afford a car. Moreover, only the city offers a large supply of jobs to the unskilled. In fact, there is a clear correlation between the number of building permits and housing prices. Even in Chicago, building land is scarce, but between 2002 and 2008, 68,000 building permits were issued here, while Boston only approved the construction of 8500 new buildings in the same period (Glaeser 2011a, pp. 150–151, 184, 242). In 2012, the average purchase price for a housing unit

◘ **Fig. 3.7** Greenwich Village Historic District. (Adapted from ► www.nyc.gov/landmarks)

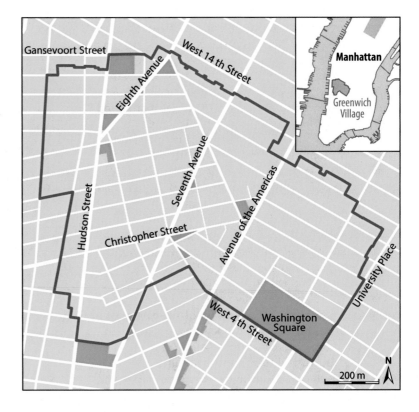

was \$462,500 in New York, \$370,000 in Boston, and only \$162,900 in Chicago (▶ www.zillow.com). Against this background, Glaeser's criticism of the restrictive granting of building permits in New York seems to be justified, even if a European can hardly imagine an even more compact development than in Manhattan. Mayor Bloomberg's answer to New York's housing crisis is to build *micro-units* (mini-apartments) of well under 30 m^2. This would not only create affordable housing, but also respond to demographic change and the large number of single households. Reactions to this proposal were mixed (New York Times 2012d, e 21/9/2012, 11/7/2012). Interesting were the considerations that one does not really need one's own kitchen in New York in view of the numerous restaurants and fast food stands. The micro-units could thus contribute to a revitalization of the sidewalks in the sense of Jane Jacobs.

Following the example of the *historic districts,* cities have designated a large number of different *special districts* in recent decades. There is hardly a city center left without an *art district*. New York has also designated the Theatre District, the Special Fifth Avenue District, the Special Garment Center District, and the Special Little Italy District, among others. In San Francisco, there are 16 different *special districts* in which certain uses are excluded or encouraged in order to preserve the character of the neighborhoods. Among planners, the *special districts* are controversial because these goals can also be achieved under *"normal"* zoning. The most important reason for designating *special districts* is probably to improve the marketing of the neighborhoods. Another way to preserve evolved structures is through the *National Register of Historic Places* (NRHP), established in 1966 to preserve America's archaeological, historical, architectural, and cultural heritage (▣ Fig. 3.8).

▣ **Fig. 3.8** Urban preservation east of Union Station in Washington, DC

Today, the register includes more than 75,000 buildings and monuments of various types. Since demolition or alteration of protected properties is almost impossible, the NRHP makes an important contribution to the preservation of older or special buildings. However, towns often feel that their planning powers are restricted (Cullingworth and Caves 2008, pp. 109–110, 224–226).

3.6 Transport and Traffic

In the USA, the automobile embodies the longing for individualism, freedom and independence. Far earlier than in Europe, the middle class was able to realize the American dream of unrestricted mobility. At the same time, public transport began to decline (Moen 2004). At the beginning of the twentieth century, many cities had a good and growing streetcar network, but with the rise of the automobile, the number of public transport users soon declined. The ultimate end of good mass transit came in the early 1930s when a company called National City Lines, formed jointly by General Motors, Firestone Tire, Philips Petroleum, and Standard Oil of California, bought up the streetcars in Los Angeles and 44 other cities, gradually shutting them down and replacing them with buses produced by General Motors. From today's perspective, the action was illegal, but at the time no one was advocating for the preservation of streetcars, which were preserved in only a few locations, such as New Orleans. Encouraged by the major automobile manufacturers, American cities were evolving into automobile-friendly cities. At the 1939 World's Fair in New York, General Motors presented a vision of the city of the future called *"Futurama"*. In the model, pedestrian and automobile traffic were separated and all major locations were connected by a network of highways. It was supposed to be possible to reach every place in the shortest possible time in one's own vehicle (Gallagher 2010, p. 73).

For decades, the car-oriented city became the most important guiding principle of planners and politicians. With government support, inner-city roads and highways were built and urban sprawl and suburbanization were promoted. In many places, inner cities in particular have suffered from road construction. Multilane ring highways were built, separating city centers from neighboring neighborhoods, or expressways were built along the shores of lakes or rivers, sometimes even elevated, as in Seattle or San Francisco. Drivers were afforded a good view of the bodies of water, but access to a recreational shoreline was denied to city dwellers. Detroit, the motor city, was probably more than any other American city converted to individual transport, but even New York was soon criss-crossed by a network of highways. Robert Moses was in charge of road construction in New York from the 1930s until 1968. Although he never held a driver's license, he believed unconditionally in the car-friendly city (Moses 1962). Under his leadership, they realized 13 bridges and 670 km of expressways, called parkways in New York, which earned him the title *"highwayman"* (Fitch 1993, p. 68). While the newer cities in the West and South of the USA expanded into the area with the construction of the roads, in the older cities on the East Coast and in the Midwest many building blocks and dwellings were demolished for road construction (Fitch 1993; New York Times 1981 30/7/1981).

The compact city of New York was able to maintain its vibrancy despite the construction of many streets, while in smaller cities the centers were often sacrificed in order to achieve optimal traffic routing. In San Bernardino, California (2010: population 210,000), several buildings on historic Main Street were demolished in the late 1960s to make way for a faceless shopping center. At the same time, the parallel street was widened to allow for better traffic flow. Several mega-complexes were connected by enclosed pedestrian bridges and the street space was given over to moving and stationary traffic. Since activities that bring people together, such as pedestrian traffic and small-scale retail, no longer existed at the street level, all urban life became extinct. Meanwhile, the negative effects of the separation of functions were recognized and a paradigm shift was initiated. Not the car-friendly city, but the livable city is in the foreground of urban planning considerations. On the edge of downtown San Francisco, the two-story Embarcadero Freeway, which collapsed during an earthquake in 1989, was not rebuilt. In Portland, OR, and Boston, downtown urban freeways were demolished and placed underground, and after much debate, Seattle is currently tunneling 3.2 km of the two-story Alaskan Way (State Route 99), which opened in 1953 and separates downtown from Elliott Bay. Completion is scheduled for 2015. In addition, many small measures were taken to make cities more attractive. In San Bernardino, realizing that the radical transformation of the city was a mistake, they turned a large parking lot in a central location into a public park that was immediately well received and is even used for weddings. At the same time, other measures are being implemented to make the city center more attractive, including the upgrading of bus stops (Gratz and Mintz 1998, p. 114). In New York, Times Square, Harold Square and Wall Street have been partially closed to traffic. During the summer months, chairs now invite people to linger here.

3.6.1 Disadvantages of Individual Transport

In view of rising gasoline prices, Americans are increasingly thinking about the economic disadvantages of individual transport. In 2009, around 15% of all private household spending was on transport and traffic (U.S. Census Bureau 2011b, Table 684). As gasoline costs have risen sharply in recent years, the share is likely much higher today. The median time to travel to work by private vehicle is just over 25 min. Each commuter spends 34 h a year stuck in traffic, and congested roads cost them $750 annually (Texas Transportation Institute 2011). However, congestion can also be viewed positively, as drivers who spend a lot of time in traffic jams every day eventually become so annoyed that they move closer to work or look for other transport options (Owen 2009, p. 139). Yet, unchanged since 1990, only about 5% use public transport to get to work, but it takes them just under 48 min. In New York MA, 30.5% of all working people use public transit, and in Chicago, Washington, D.C., San Francisco, and Boston MAs, which also have good mass transit networks, between 11% and 15% each. Because it is not feasible to establish efficient mass transit in very sparsely populated urban regions, less than 5% of the workforce uses public transit in Houston and Atlanta (McKenzie 2010). In cities with low population density, it is almost impossible to run errands on foot, while in compact cities a growing number of people use their own legs as a means of transport. In central New York or Chicago, there is good public transport, but it is not worth waiting for a bus or subway to cover short distances. Besides, there is plenty to see on the lively streets of these cities. The New Yorker lives nine months longer than the average American; possibly because he uses his own locomotor system more often than his compatriots (Owen 2009, pp. 163–168).

Cars are a status symbol and their constant availability is their greatest advantage. In cities with hot and humid climates, the car also serves as an air conditioning unit between two air-conditioned buildings (Owen 2009, p. 166). On a positive note, the use of shared cars has increased (◻ Fig. 3.9). Car sharing was introduced in the U.S. in the late 1980s. In 2010, ZipCar, by far the largest provider, already provided 9000 vehicles at 4400 locations, which could be rented by ZipCar's 400,000 members (New York Times 2010 10/9/2010). Car sharing is only profitable for providers in densely populated locations and for customers only for short trips. Overall, there is a lot to be said for owning a car in the USA. The offer and image of public transport are very poor in many locations. There are often no bus schedules, and where they do exist, they are not adhered to. For the

Fig. 3.9 Parking space for *ZipCars*

Fig. 3.10 High parking fees in Manhattan

poor population it is therefore hardly possible to pursue a regular occupation if they cannot reach the workplace on foot. However, the unskilled in particular often work at times when buses and trains rarely run (Lyons 2011, pp. 294–295; Moen 2004).

3.6.2 Location of Residence and Choice of Transport Mode

Ownership of a private car determines the location of residence. In suburban areas, a family can only live if the household has at least one, but preferably two or three cars. Access to public transport affects segregation in different ways in older and newer cities. In New York, Boston, and Philadelphia, four areas can be distinguished based on residents' income and their transport choices. The affluent who live in the central sub-areas of the cities, such as Manhattan or in Beacon Hill in Boston, walk, take taxis, or use public transport (■ Fig. 3.10). On the edges of the core cities, such as the boroughs outside Manhattan or Roxbury in Boston, live the poor who travel long distances by public transit. In the adjoining counties like Westchester County west of New York, the wealthy, who usually have mul-

tiple vehicles, live on large lots. At even greater distances from the core city, lots are less expensive and affordable to the middle class, but they rely on cars in this location. The newer cities like Los Angeles lack the inner zone where many pedestrians can be found. Here the affluent use cars everywhere and only three zones can be distinguished. In central city areas such as South Central, the poor rely on public transit; in affluent neighborhoods such as Beverly Hills, the car is used; and middle-income people living in the adjoining counties face long commutes (Glaeser 2011a, pp. 85–86).

It is encouraging that public transport has been expanded in many cities in recent years. In Cleveland, for the first time in the U.S., a bus rapid transit line (BRT) opened on Euclid Avenue, connecting the downtown arts and culture center with the Cleveland Clinic 10 km away. Several new apartment complexes were built along the new route during construction, and more are planned. The so-called Health Line is making an important contribution to the revitalization of the city (The American Assembly 2011, p. 10). A similar effect is expected when the Eastside Subway opens in Manhattan. Currently, the tunneling of Second Avenue is causing great inconvenience to residents and flowing traffic. Once the new subway line is completed, though it

3

won't be until late 2016 at the earliest, the East Side will become more attractive. A number of American cities, such as Phoenix and Seattle, have revived or rebuilt old streetcar lines, known as light rails in the US, in recent decades. Success, however, is not always guaranteed. In Silicon Valley, the first stations of a light rail called VTA (Valley Transit Authority) opened in 1987 and now serves 62 stations over a length of nearly 68 km. Construction has so far cost $2 billion. Although free Wi-Fi is even available on the trains in tech-savvy Silicon Valley, the VTA is hardly used. On its 25th anniversary, it was reported that annual operating costs are $66 million, 30% higher than other cities. At the same time, the VTA's utilization rate is 30% lower (Associated Press 2012b, 12/27/2012). The long distances and the fact that free parking is available in large numbers in Silicon Valley are probably important reasons for the low acceptance of the VTA.

3.7 Green Cities for America

U.S. cities are considered to be environmentally hostile. With low population density, the degree of sealing is high, large areas are taken up by moving and stationary traffic, public transport is poor and dependence on private cars is high. The houses are hardly insulated and consume a lot of energy for heating or air conditioning, and huge wastelands disfigure the cityscape.

Despite all these shortcomings, compact cities are more environmentally friendly than those with only a low population density or suburban space, and several scientists surprisingly highlight the city of New York as a particularly positive example. Since the living space per inhabitant is small and hardly any private cars are used, the ecological footprint of a New Yorker is comparatively small. Manhattan residents today consume no more gasoline than the average American in the mid-1920s. Every single New Yorker consumes less electricity and generates less waste than residents in the surrounding areas. Even the high-rises are more environmentally friendly than many suburban office buildings. The photovoltaic panels on the façade of the 48-story Condé Nast building in Times Square, completed in 1999, have less benefit than the uninformed observer might suspect, generating barely 1% of the building's electricity consumption; but even if many of the supposedly eco-friendly elements are more show than reality, the building's sheer size saves the environment. The skyscraper occupies an area of 0.4 ha and has a floor area of 150,000 m² distributed over 48 floors. If the same floor area were to be created in single-storey buildings in a suburban *office parc* and equipped with the necessary parking spaces and the usual green areas, at least 60 ha of land would be required. There is no doubt that cit-

ies with large population densities have low individual pollutant emissions. Nevertheless, it is unfair to call New York the most environmentally friendly city in the United States, because the metropolis would not be able to survive without the emission-polluting industry and agriculture in other regions of the country (Owen 2009, pp. 44, 205, 213).

In Boston and Philadelphia, there are also large differences between the core city and the suburban area due to the good public transport and the high development densities in the central city areas. In the metropolitan areas of Atlanta and Nashville, there are hardly any differences between the city and the surrounding area. The cities of Atlanta, Dallas, Memphis, Oklahoma City and Houston, located in the southern US, have the highest CO_2 emission values, as they generate the energy for the frequently used air conditioning systems from coal. In the sprawling urban landscapes, people are also particularly likely to drive their own cars. In these cities, CO_2 emissions per household are nearly twice as high as in many California cities. A comparison of emissions in 48 metropolitan areas showed that San Francisco, San Jose, San Diego, Los Angeles and Sacramento have the lowest CO_2 emissions per household. Because of California's moderate climate, spending on heating or air conditioning is relatively low here. In addition, environmentalists have enforced that only energy-saving household appliances may be used, and much electricity is used from the environmentally friendly energy sources of water and gas. California is now considered one of the most environmentally friendly states in the USA, but here too appearances are often deceptive (Glaeser 2009, 2011a, p. 14; Owen 2009, pp. 2–7, 17; Schulz 2008).

In recent times, many cities have begun to rethink their approach, as the problems of a laissez-faire attitude towards the environment are increasingly recognized (◘ Fig. 3.11). Antonio Villaraigosa announced in

◘ **Fig. 3.11** Innovative waste bin in San Francisco

2005, when he took office as mayor of Los Angeles, that his city should become the greenest major city in the USA. He created new agencies such as the Department of Water and Power and the Harbor Department and gave them broad powers. He has also adopted a plan to reduce CO_2 emissions by 30% below 1990 levels by 2030. In 2007, New York Mayor Bloomberg released the ambitious *"Plan NYC: A Greener, Greater New York"* to prepare the city for another million residents, strengthen the economy, mitigate climate change, and improve the quality of life for all residents. Twenty-five city agencies are working together to achieve this goal, which includes some 130 projects such as protecting wetlands in Jamaica Bay, expanding public transit, and reducing carbon emissions (Halle and Beveridge 2011, p. 161). A showcase project for efforts to contribute to climate improvement is the 366 m high Bank of America office building not far from Times Square, completed in 2009, which generates two-thirds of its own energy consumption. The floor-to-ceiling glass windows of the offices are optimally insulated and guarantee working in daylight deep inside the building. The skyscraper was the first commercial building in the USA to be awarded the Platinum Certificate of the U.S. Green Building Council (Lamster 2011, p. 83).

Comprehensive plans have the disadvantage that they are very expensive and sometimes take decades to implement. New local politicians may pursue other goals or economic crises may make people forget their concerns about the environment. Often small projects initiated by local authorities, citizens' initiatives or churches are much more sustainable because they combine environmental protection with an improvement in the quality of life. Such projects include the renaturation of streams or the creation of Highline Park in New York and *community gardens* at various locations in the USA. In the past, people threw their waste into ponds or rivers, which not infrequently became stinking trickles. The best way to solve this problem was to channelize and cover the water bodies. The stench disappeared, but it was not uncommon for residents to suffer from wet cellars from then on. In the meantime, the aesthetic value of the watercourses has been recognized and they have begun to be partially brought back to daylight. This process is called *daylighting*. In Kalamazoo (pop. 74,000) in southern Michigan, Arcadia Creek flowed underground for about 100 years until it was renaturalized along a five-block length downtown in the 1990s. Several ponds were added. Today, Arcadia Creek, the ponds, and the surrounding meadows provide effective flood protection. In addition, waterfront properties have increased in attractiveness and value. Similar measures have been implemented in Oklahoma City, Kansas City, and Indianapolis (Biello 2011; Gallagher 2010, pp. 90–96; Hamilton County Planning and Development 2011).

3.7.1 High Line Park in Manhattan

In the early 1930s, an elevated railroad had been built in Chelsea on the west side of Manhattan between 34th and Spring St. to shift the transport of hazardous materials from the industrial and port facilities in Lower Manhattan, which had previously been transported via 10th Avenue. With the decline of the port's importance, the elevated railroad was used less and less, and service was discontinued in 1980. The alignment was demolished between 16th and Spring St., while the remaining portions rotted and became overgrown. Two environmental activists first advocated for its conversion into a public park in 1999. In Paris, the first section of the Promenade Planté had just opened at that time. As in Paris, they wanted to create a public park on the old railway line in New York. The idea was quickly taken up in the neighborhood and the citizens' initiative *"Friends of the Park"* was founded. However, at that time, some of the residents would have preferred the unsightly wasteland to be torn down. Thanks to good contacts with politicians and other important personalities in New York, the activists succeeded in 2001 in making the conversion of the old railway line an important issue in the election campaign for the office of New York mayor. Eager to win votes, all six candidates endorsed the new park (◉ Fig. 3.12). Opponents of the project, however, had a key ally in incumbent Mayor Rudolph Giuliani (1994–2001), who signed a document in the last week of his term that determined that the site would be demolished. Friends of the Park successfully challenged the order in court, but although Mayor Bloomberg had voiced support for the park's construction during the election campaign, he declared in early 2002 that the city's financial burdens made it impossible to implement the project. This statement, in turn, galvanized activists, who projected that the $65 million cost of the transformation

◉ **Fig. 3.12** High Line Park in New York

would be recouped in just 20 years by more than doubling the profits. Mayor Bloomberg eventually had to cave to public pressure. After further squabbling between the responsible authorities and changes in the zoning plan, construction finally began in 2007. Since the press had reported extensively on the planned park for years, the prices for the land located along the railroad line rose sharply starting in 2008. Investors assumed that there would be a lot of interest in an apartment overlooking the new park. The first section of High Line Park opened in June 2009 and was an immediate success. A second section opened in the summer of 2011. The park stretches for about 2.3 km between Gansevoort and 30th St.; an extension to 34th St. is planned. Extremely popular with tourists and locals alike, the former railroad right-of-way runs partly through ravines with houses of all ages and states of maintenance, giving it a surreal feel. Each year, a reported four million people visit High Line Park. It is obvious that the shops in the Meatpacking District, which is located at the southern end of the railway line and has in any case been *gentrified* in recent years, benefit from the many visitors. High Line Park now provides a green lung on the densely populated Lower West Side and enhances the neighborhood. In 2012, the average price of an apartment was just under $1.4 million, surpassed only in a few locations in New York (Halle and Beveridge 2011, pp. 162–166; ► www.highline.org; ► www.zillow.com).

3.7.2 Community Gardens

Already at the end of the nineteenth century, home cultivation had been propagated in some cities such as New York, Boston, Chicago and Detroit, and in 1891 a school in Boston received the first school garden in the country. Especially in times of crisis, such as after the First World War or during the depression at the beginning of the 1930s, many kitchen gardens were planted in cities, and during the world wars *victory gardens* helped to ensure the food supply of the urban population (Sokolovsky 2010, p. 245). In New York, *community gardens* or *urban gardens* have made an important contribution to the nutrition and social stabilization of the population in troubled neighborhoods since the 1970s (Grünsteidel and Schneider-Sliwa 1999). The desire to lead a more conscious and simpler life, as well as to protect the environment by avoiding excessive fertilizing and the transport of food over long distances, has led to an intensification of the movement since the 1990s. Land for which there is no demand and is therefore worthless is put to temporary use, which is by no means intended to be permanent (Bishop and Williams 2012, p. 65). The

movement became known beyond the borders of the USA when Michelle Obama planted vegetable beds in the White House garden with children from disadvantaged neighbourhoods in 2009.

In many cities, the supply of food in the neighbourhoods of the poorer population is very poor. As profit expectations are low and expenditure on security is high, the large chains that dominate food retailing do not invest in these neighbourhoods, and *food deserts* are created (Acker 2010, p. 137). The poor population often does not have a car, and public transport, if available, is expensive and unreliable. In order to guarantee a minimum supply for the population, food stamps may be redeemed at petrol stations in many places, but their supply is usually limited to potato chips and similar high-calorie and low-nutrient snacks. It is estimated that approximately 80% of Detroit residents obtain their food from stores with a limited and poor supply or from gas stations (Renn 2009). It is not surprising that many people in Detroit and comparable cities or neighborhoods are obese and deficient. In Detroit, arguably the largest *food desert* in the United States, activist Grace Lee Boggs initiated the *Gardening Angels* movement around 1990, which was soon followed by similar initiatives in other cities. As a rule, the vegetable and fruit gardens are run by a single family or in the form of community gardens. However, *urban gardens* are far from uncontroversial. Often, residents seize plots of land without any permission. The legal relations are difficult, because the land is either owned by an (absentee) owner or by the city. Municipalities do not want to sell the land lest they permanently deprive it of another use. In New York, Mayor Rudolph Giuliani wanted to destroy more than 100 community gardens for housing in the late 1990s. Protests against the mayor and the investors only died down when the actress Bette Midler bought the plots of land shortly before the planned destruction of the gardens and handed them over to the community gardeners. This example is not unique to New York, where building land is expensive and protests or court appointments over wrongful seizure of brownfields are commonplace. There are about 500 *urban gardens in* New York and also about 400 gardens maintained by seniors living in the city's public housing. The gardens not only serve the basic needs of the urban population, but also have great therapeutic value. They have even been shown to reduce crime rates (Gallagher 2010, pp. 45, 53–54; Sokolovsky 2010, pp. 244–249).

Today, a significant number of *community gardens* are planted by non-profit organizations such as Growing Power in Milwaukee, Las Parcelas in Philadelphia, and Earthworks in Detroit (◘ Fig. 3.13). In Detroit, the friars of a Franciscan order opened a soup kitchen

Fig. 3.13 Earthworks Community Gardens in Detroit

for the needy in 1997, growing vegetables to feed them on a vacant lot near their convent. Over the years, the community took over more and more fallow land and built a large greenhouse in 2004. Today, a few permanent staff and many volunteers produce a wide range of food and even their own honey. The produce has long exceeded the demand of the soup kitchen and is sold on the Earthworks site to the Detroit community. Cooking classes for adults and children are also offered. A major problem is the city's potentially polluted soils. Some of the production therefore takes place in raised beds, some of which are filled with soil from a religious community cemetery. Modeled after Earthworks, the Garden Resource Network has sprung up in Detroit, overseeing some 1400 private gardens, *community gardens*, and school gardens. There is no doubt that *urban gardens* make a positive contribution to feeding Detroit's population. Unfortunately, despite all efforts, only a few young and unemployed can be persuaded to dig in the soil in the summer heat. It is this segment of the population that could be introduced to regular work. Many gardens produce large surpluses at times, which they sell on Saturdays at the Eastern Market near downtown Detroit, which is also popular with suburban residents. The value of the gardens can hardly be overestimated, as they significantly improve the food situation, green up fallow land and enhance the cityscape (Gallagher 2010, pp. 47–72; information from Earthworks staff).

Reurbanisation and Restructuring

Contents

In recent years, some U.S. cities have experienced an upswing, which in America is unspecifically referred to as *revitalization*, in German as Revitalisierung or more precisely as Reurbanisierung. There are different definitions for revitalization as well as for reurbanization, describing either quantitative or qualitative processes, where the prefix "Re" announces a reversal of a longer lasting process, which however does not have to lead to the restoration of a former state (Brake and Urbanczyk 2012, p. 35). Depending on the question, reurbanisation can be identified using quantitative or qualitative indicators. Migration flows, an increase or decrease in population or jobs can be measured precisely. It is possible that these increase again in absolute terms in the core city after years of negative development or that they increase more strongly in relative terms than in the suburban area. According to the 2010 census, at the beginning of the new century, some of the core cities experienced large population losses, while others experienced population gains, though always far below the growth rates of the suburban area. The suburban area far from the city made the highest gains. Although the core cities have experienced a loss of importance compared to the surrounding areas, one can speak of reurbanization if the absolute number of inhabitants has increased. Based on observations first on quantitative reurbanization in European densely populated areas, where these processes started much earlier than in the US, models of urban development have been developed that distinguish the phases of urbanization, suburbanization, desurbanization and reurbanization, although the chronological order is not mandatory (Gaebe 2004, pp. 18–157). In U.S. cities, an overlap or even simultaneity of the individual phases can be observed. This also applies to suburbanisation and reurbanisation, which may well occur at the same time.

In recent times, the focus of attention has been far more frequently on qualitative changes in the sense of reurbanisation as a series of upgrading processes in the inner cities or areas close to the inner cities, which are accompanied by social and economic changes and have a positive impact on the image of the city as a whole. The global competition to which cities are exposed and liberal policies offer an important explanatory approach to this process, which is associated with a restructuring of public space (Sorkin 1992). Visible signs of reurbanization such as flagship stores, expensive shopping malls and restaurants, or luxury apartment complexes can be found especially in inner cities. *Gentrification* is a specific manifestation of reurbanization (Brake and Herfert 2012, p. 16; Gerhard 2012, p. 54). The new urbanites and affluent tourists have contributed to the emergence of consumer cities. Qualitative reurbanization may be limited to selected parts of the city, while at the same time other parts are (still) in the downturn phase (Brake and Urbanczyk 2012, p. 35).

4.1 Global Cities

U.S. cities are involved in global competition, with New York undoubtedly at the top of the hierarchy, followed by Chicago, Los Angeles, San Francisco, Washington or Miami. The development of these cities has not been linear over the past decades; phases of a relative gain in importance have alternated with phases of a decline in importance (Fainstein 2011). The number of headquarters of transnationally active companies represents an important indicator of the significance of a *global city*, as this is where the important decisions are made and the world-spanning company networks are coordinated (Kujath 2005, p. 13). With 18 headquarters of the 100 companies with the highest turnover in the USA, New York is the undisputed leader. These are all concentrated in just a few blocks south of Central Park and at the southern tip of Manhattan. They are followed by San Francisco with eight, Chicago with six, Minneapolis with five, and Houston, Atlanta, and Washington, D.C., with four corporate headquarters each (◘ Fig. 4.1). In the locations mentioned, however, the headquarters are not concentrated in such a confined space as in New York, but are in some cases located outside the core cities. This is particularly true of San Francisco, where several high-tech companies such as Apple or Hewlett Packard are located in Silicon Valley to the south. These are therefore *global regions* rather than *global cities*. *Global-city* research has predominantly focused on New York, Chicago and Los Angeles, but Washington, D.C., also represents an interesting case study.

4.1.1 New York

In the seventeenth century, the colonists chose the location of the later city of New York because of its proximity to the Atlantic Ocean as well as its sheltered location, which was usually ice-free in winter. In addition, the Hudson River provided access to a large hinterland. Since trade was not limited to the mother country and other colonies early on, New York developed into the largest port in the world by 1820. It experienced an enormous increase in importance with the opening of the Erie Canal in 1825, which connected the Hudson River with the Great Lakes. Soon the railroads opened up an ever-expanding hinterland. New York became the hub of national and international trade in goods and, due to the great demand for labor, the most important port of call for immigrants, who guaranteed an always large supply of cheap labor (Abu-Lughod 1999, pp. 37, 72; Jackson 1995, pp. 581 f., 926 f., 977 f.). In the nineteenth century, New York developed into an industrial city in which the clothing industry, the steel industry and the printing and publishing industries became estab-

4

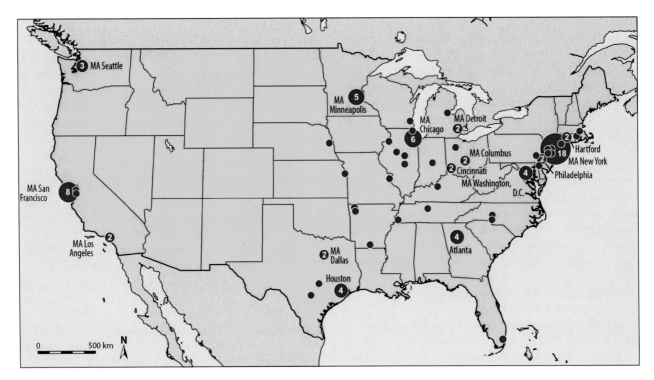

■ **Fig. 4.1**   Headquarters of the 100 companies with the highest turnover in the USA in 2012. (Data basis ▶ www.fortune.com)

■ **Fig. 4.2**   New building of the New York Times, one of the world's most important newspapers

lished. Regular shipping connections to Europe guaranteed the reliable receipt of news from the Old Continent, and New York developed early on into the country's leading press and publishing center (■ Fig. 4.2). Huge fortunes were made from the expansion of industry, the transport system, and commerce, as well as from land speculation. Its rise as the nation's largest banking center was almost inevitable. In 1817 the New York Stock Exchange was founded, and in 1865 it was moved to its present location on Wall Street, where the Bank of New York had already established itself in 1797. Soon

the city rose to become the dominant financial center of the country (Jackson 1995, pp. 74, 261 ff., 810, 844 ff.). The large fortunes were crucial to New York's development as the cultural capital of the USA. Founders and investors of museums, opera houses, concert halls and libraries that would later achieve world fame were, for example, the land speculator J. J. Astor and the banker J. P. Morgan, who was a co-founder of the Metropolitan Museum (Homberger 1994). In the twentieth century, industry and commerce initially remained significant, the banking sector expanded, and the port remained the largest in the world until it was overtaken by Rotterdam in 1960. With the stock market crash of 1929, the optimistic mood changed abruptly. A slight upturn emerged towards the end of the 1930s, but it was not until the war in Europe that the final turning point came (Abu-Lughod 1999, pp. 37, 185–191).

Due to suburbanization and deindustrialization, the number of industrial jobs has steadily declined since the late 1960s (City of New York 2002, p. 2 f.). In the 1970s, New York went through a major crisis. The infrastructure was outdated and the port was of little importance. Although taxes were among the highest in the U.S., the city was temporarily insolvent. In 1975, bankruptcy was averted only by the Municipal Assistance Corporation taking over its finances. The population declined during that decade, but by the 1980s an economic upswing had begun, which was reflected in rising population numbers. The textile workers had educated their children well and learned to take risks. New York was able to expand its function as a *global city*, attracting and managing international capital like no other city in the world (Glaeser 2011a, pp. 3, 56–57; Sassen 1993). The largest American banks, brokers and insurance companies as well as the most important foreign banks have their headquarters here. Along with Amsterdam, 47th Street is the leading trading center for diamonds. With the headquarters of the United Nations located on the East River, New York is also the scene of world political decisions. Although hardly any textiles are produced in New York today, the fashion sector is still vibrant. Many designers such as Donna Karan and Calvin Klein live and work in New York, but their designs are turned into finished garments on the other side of the world. New York's museums can be compared to those of London or Paris, and the private art market attracts collectors from all over the world. When it comes to plays, musicals and music, New York impresses with its abundance of offerings. The city is also a global shopper's paradise, with even Europeans jetting off for Christmas shopping. However, New York is a city of contrasts, and rich and poor live side by side in close quarters. While the Bronx and northern Manhattan are partly characterized by great poverty, wealth and luxury dominate in parts of southern Manhattan (❑ Fig. 4.3). Between 41st and 59th St., 600,000 people now work, more than in the states of New Hampshire or Maine, earning more than $100,000 a year on average (Glaeser 2011a, p. 5; Hahn 2003). In the 1990s, the city benefited from the rise of the high-tech sector and global financial markets. Disillusionment set in with the bursting of the Internet bubble, the crash of stock market prices in March 2000 and the terrorist attacks in September 2001. The banking crisis and the bankruptcy of the investment bank Lehman Brothers in autumn 2008 gave the city another shock. But New York has long since recovered from

this crisis as well. The number of inhabitants is increasing, and real estate prices and managers' bonuses have long been on the rise again (New York Times 2012a 4/11/2012).

4.1.2 Chicago

Chicago was founded in 1833 with only about 350 residents. The city owes its rapid growth and rise to a hub for the movement of goods to its location on Lake Michigan, which is connected to New York via the other Great Lakes and the Erie Canal. In addition, since 1848, the Illinois and Michigan Canal has connected the city to the Mississippi River and thus to the Ohio River and the Gulf of Mexico. In the 1850s, the port developed into the world's largest trans-shipment center for lumber and grain. By 1857, Chicago was the largest railroad hub in the world, with rails converging at nearly 5000 km (Miller 1997, pp. 89–121). Its rise as a major stock exchange was closely linked to the grain trade. In the 1850s, the Chicago Board of Trade allowed for the future buying and selling of grain at a predetermined price. The system of *future markets*, which was soon extended to other products, resembled a bet on the expected harvest or demand. Meat production became of paramount importance. Livestock raised in the western United States were transported to Chicago by rail, and the Union Stockyards, opened in 1865, soon became the largest slaughterhouses in the world (Abu-Lughod 1999, pp. 100–116).

Numerous innovative entrepreneurs promoted the boom to a cosmopolitan city. McGormick developed new types of agricultural machinery and Pullman built the saloon car. The invention of the refrigerated truck in Chicago in 1869 was important, as meat produced in the city could now be delivered as far as the East Coast. In 1893, slaughterhouses processed 14 million animals. Steel production boomed in the 1880s due to the expansion of the railroad network. The large number of immigrants, mostly from Eastern Europe, guaranteed the city a steady influx of cheap labor. By 1890, Chicago had become the second largest industrial city in the United States after New York and had 1.1 million inhabitants only 60 years after the city's founding (Mayer and Wade 1969, pp. 3–192; Miller 1997; Miller 1997, pp. 24–253). Tremendous urban growth and industrial expansion continued in the first decades of the twentieth century, but the depression of the early 1930s hit Chicago far harder than New York. After the end of the war, the economy initially stabilized, but entered a prolonged crisis in the late 1950s. Because the U.S. concentrated defense industry contracts in coastal locations during the Cold War, Chicago did not benefit to the same

❑ **Fig. 4.3**   Military parade on the world-famous 5th Avenue

4

☐ **Fig. 4.4**   Boeing headquarters in downtown Chicago

☐ **Fig. 4.5**   Universal Film Studios

degree as the New York/Connecticut and Los Angeles regions (Abu-Lughod 1999, pp. 212–236). Even though Chicago has been only the third largest city in the US since the 1980s, it is the second most important stock exchange in the U.S. after New York and a major financial center. The McCormick Place convention center is one of the largest in the country. Chicago's image was boosted enormously when the aircraft manufacturer Boeing relocated its headquarters here from Seattle in 2001, even though the city made the move possible in the first place by providing large subsidies (☐ Fig. 4.4). However, only two of the top 100 companies, Boeing (ranked 39) and United Continental Holdings (ranked 76), have their headquarters in Chicago. Four other companies in the top 500 are located in the core city. Another five headquarters of the top 100 and 17 headquarters of the top 500 are located in the surrounding area, including such household names as Kraft Foods (ranked 50), Sears (ranked 65), and McDonald's (ranked 107) (as of 2012).

4.1.3 Los Angeles

Los Angeles grew out of a mission station founded by the Spanish in the eighteenth century. Although its location on the Pacific Ocean away from Europe in a semiarid region and the lack of a natural harbor precluded any major growth, the California gold rush of the 1840s, the connection to the continental railroad in 1876, and the oil boom that began in 1892 led to a boom. Beginning in 1924, the port benefited from the opening of the Panama Canal, and in 1916 it was designated a base for the Pacific Fleet. During the Great War, the United States also encouraged shipbuilding at the site. Inexpensive energy from petroleum and hydroelectric power benefited the expansion of private industry.

Los Angeles developed into the center of the oil industry, the second largest tire producer in the country, the most important furniture, glass, and steel producer in the West, and the regional center of automobile and aircraft manufacturing as well as chemicals (Abu-Lughod 1999, pp. 133–163, 237–268). At the beginning of the nineteenth century, the center of the global film industry emerged here. This had initially settled in Chicago, but was soon moved to Los Angeles due to better weather, where the first film was shown to the public in 1902. By 1919, 80% of all films produced worldwide were made in California. Production companies such as Paramount, Universal, Fox Film and Loews-Metro-Goldwyn-Mayer marketed their films worldwide (Doel 1999) (☐ Fig. 4.5). In the following decades, Los Angeles, similar to the Bay Area centered on San Francisco, Seattle, and San Diego, benefited from America's entry into the Pacific War in late 1941, the Vietnam War (1964–1973), and the Korean War (1950–1953) due to the associated military investments. Until World War II, Los Angeles had produced mainly for the local market, but from the 1950s it increasingly produced for the world, benefiting in particular from its Asian counterpart. However, with Walt Disney and Ingram Micro, only two headquarters of the largest American companies are located in Los Angeles MA. Although other global centers such as India's Bollywood have developed in the meantime, Los Angeles' production companies still dominate the film industry of the Western world and transport California's image to the farthest corners.

4.1.4 Washington, DC

The American capital has long been neglected in *global city* research, although the city on the Potomac River is the undisputed center of world politics (☐ Fig. 4.6).

☐ **Fig. 4.6** The Capitol in Washington, DC

However, an orientation towards purely economic factors seems one-sided (Gerhard 2003, p. 57). In 1790, the location of the new capital of the United States of America, founded in 1776, was determined to be about 150 km west of the Atlantic coast. The location was chosen for strategic reasons on the border of the quarreling southern and northern states, and the city area was carved out of the states of Maryland and Virginia. The Frenchman Pierre Charles L'Enfant was commissioned to lay out the city, which was based on the plans of European absolutist baroque cities with wide boulevards, central axes and representative squares. However, it took about 100 years for the plan to be realized with some modifications. The large memorials such as the Lincoln Memorial and the Jefferson Memorial were even not completed until 1927 and 1942 respectively (Holzner 1992).

In recent decades, the MA Washington (2010: 5.6 million inhabitants) has developed very positively and is comparable in economic terms with similarly large metropolitan areas. Washington, D.C., is of outstanding importance as a *global city* due to the high institutional density of political actors. With few exceptions, all states have one of their largest embassies here, and the World Bank and International Monetary Fund are located just a few blocks from the White House on Pennsylvania Avenue. In addition, more than 8000 nongovernmental organizations (NGOs) of varying sizes and some 300 think tanks and foundations operate in Washington, D.C.. Many of the NGOs are globally connected via the Internet and deal with international issues. In Washington, D.C., itself, NGOs and other internationally active institutions are concentrated almost exclusively in the northwest sector of the city, where the White House is located. This part of the city, as well as the adjacent suburban area, is also the preferred residential location for the largely highly qualified staff of interna-

tional organizations from around the world, where they have contributed to the *gentrification* of once decaying neighborhoods like Adams Morgan and the exorbitant rise in land prices. There are no ties to the city's poor in the neighborhoods south of the Anacostia River, although many NGOs are involved in fighting poverty in Africa or other faraway continents. Like many other *global cities*, the US capital is a highly segregated city (Gerhard 2003, 2007, pp. 171–188, 190–255).

4.1.5 Outlook

Despite all their differences, New York, Los Angeles, Chicago and Washington, D.C., are characterized by many common features. These include the enormous expansion of urban space and the ongoing suburbanization, the economic supremacy in the U.S. and the heterogeneity of the population. But there are also many differences such as the concentration of the financial sector in New York, the film industry in Los Angeles, the commodity futures exchanges in Chicago, and the high density of political institutions in Washington, D.C.. Globalization has neither eliminated the differences between cities nor helped to unify them (Mollenkopf 2011, pp. 174–175). In conclusion, it is important to emphasize that it is not only the major *global cities* that are affected by the worldwide restructuring of the economy, but all U.S. cities. Moreover, the process of globalization is by no means complete and its further development is hardly foreseeable.

4.2 Smart Cities

New means of communication and technologies, like other innovations before them (e.g. railways, telegraphy, mass transport), will affect the spatial organisation of the city and create new opportunities for further development (Harvey 1996, p. 412). In what ways this will occur, however, is a matter of debate. Urban geographers and planners have no experience of invisible flows, because until recently their analysis and planning was limited to visible infrastructure such as roads, railways, waterways, bridges and stations. The use of these transport routes could be accurately measured and the impact on a city's locational fabric reasonably reliably inferred. As the Internet and smartphones have taken hold in just a few years, and communication flows are almost impossible to measure and even more difficult to assess, concrete statements are nearly impossible. While some scholars predict increasing decentralization, according to other experts, cities will experience an increase in importance due to the use of new technologies. Kotkin

4

and Siegel (2000, pp. 2–7) believe that technological change will have as serious an impact on spatial development as industrialization did in the nineteenth century. Back then, railroads and mass production gave rise to large industrial centers. Today, the new means of communication are causing decentralisation. New jobs are being created on the periphery. The two authors call this development outside-in (p. 2). Since locations can be freely chosen, new technologies and urban *sprawl* are linked. Locations can be relocated almost at will, as work can be carried out indiscriminately at any point. In contrast, Saskia Sassen (2002, p. 40) does not expect decentralization through the use of new technologies, but assumes that the old established centers will expand their importance. Although the urban region will become increasingly fragmented, a loss of importance is not to be expected. Laaser and Soltwedel (2005, p. 72, 88) compared a larger number of studies on the impact of new technologies on cities and found that the spatial patterns of the old economy are more likely to be replicated by the new economy than changed. Cities have not been adversely affected by the increased use of new technologies, and in fact have been able to expand their central city functions. In the case of Chicago, it has been demonstrated that the city benefited from its convenient location on the North American continent during the industrialization period, as it does today. Many of the railroads coming from the West once converged here, and the city was connected to the East Coast via the Hudson River, the Erie Canal, and the Great Lakes. Today, while the railroads are much less important than they were in the first half of the twentieth century, Chicago benefits from the railroad network because fiber optic cables have been laid along the old routes. These converge in downtown Chicago, where they are laid in the former railroad tunnels. In 1999, the Lakeside Technology Center, which houses some 70 Internet service providers, was established in the building where the Sears mail-order catalog was once printed. Reportedly, no other location in the USA is better connected to the new technologies today (City of Chicago 2003, p. 27).

4.2.1 Influence of New Technologies

Whether the use of new technologies will trigger a decentralisation drive or strengthen the importance of cities can only be clearly clarified in the medium to long term. However, it cannot be denied that the new means of communication have led citizens, politicians, companies and public administrations to use cities differently today than they did just a few years ago. These opportunities have only recently opened up and there is no end in sight.

◨ **Fig. 4.7**   Streetcar with WI-FI in San Jose

Increasingly, city residents are connected to each other via broadband and Wi-Fi. The cost of this has become increasingly less and many cities are even planning to offer this service for free in the future. New York, whose tech-savvy Mayor Bloomberg (2002–2013) wanted all residents and visitors alike to share in the benefits of the new technology, is leading the way (Bloomberg 2011). Similar efforts are underway in Silicon Valley, where free WI-FI is offered on streetcars (◨ Fig. 4.7).

Networking is creating *smart cities*. American cities are still making too little use of the possibilities offered by modern communications to improve infrastructure or reduce costs. Following the example of London or Stockholm, access to city centers could be limited by motorized individual traffic and waste collection or water supply could be optimized. Large companies such as IBM, Cisco Systems and Siemens have recognized the potential and are working on concepts (*smart systems*) to strengthen urban infrastructure by making better use of the new means of communication (Ratti and Townsend 2011, pp. 44–45). Almost every American today has a smartphone in his or her pocket, keeping them informed of important (and unimportant) events and trends anytime and anywhere. The modern consumer does not want to miss anything. All desires must be satisfied in real time, and every craving for a product must be instantly gratifiable. Via the Internet, e-mail or Twitter, it must be possible to find out at any time and at any place where a product is available quickly and at the best possible price. ▶ Trendwatching.com refers to this behavior as "*nowism.*" Consumers are no longer called *consumers*, but *transumers* because of their ever-changing desires. More and more spaces are emerging where work and consumption can be combined. Starbucks and similar providers are the best examples of this development. The purchase of a cup of coffee guarantees free WI-FI and a warm workplace surrounded by like-minded people for several hours (Bishop and Williams 2012, pp. 68–69).

In addition, there are a large number of other services that make use of the new technical possibilities. In a joint project, the National Building Museum in Washington, D.C., Time Magazine, IBM and the Rockefeller Foundation are developing smart cities in which all socioeconomic groups of the population are to benefit from technological innovations such as WI-FI, smartphones, tablet computers and Web 2.0 (The American Assembly 2011, p. 19). Every month, different topics are discussed in discussion forums on the project's website and opinion polls are conducted (▶ www.nbm.org/intelligentcities). Apps for mobile phones not only offer almost unlimited possibilities for information, but also make it easier to meet friends. By means of the app "*Foursquare*" registered users can announce their location at any time and rate restaurants or sights. The site ▶ www.familywatchdog.us offers the possibility to check within a few seconds whether there are sex offenders in the immediate vicinity. A map not only points out the exact place of residence, but also gives the name of the criminal and clarifies the exact crime such as "*molesting children*" or "*rape.*" The website ▶ www.crimereport.com works in a similar way, but provides information on a wide range of crimes and minor offences and, where known, on the offenders. These sites can be useful when looking for an apartment, because who would want to move into a neighborhood with a particularly high number of residential burglaries or next to a multiple rapist. However, these sites are extremely controversial, as they have repeatedly pilloried innocent people and stigmatized entire neighborhoods. Of particular concern is that ▶ www.crimereport.com advertises an app with "*stay informed wherever you are,*" which allows people to stay informed about crime in their neighborhoods anytime, anywhere. However, not all services that report on happenings in the neighborhood are successful. On the website ▶ www.everyblock.com, maintained by the television station NBC, several years residents have dealt with all the developments and observations in their block. Probably never before has there been news on a comparably small scale. Anyone worldwide could participate and weigh in on the discussion boards (Ratti and Townsend 2011, p. 48). In February 2013, NBC discontinued the service without giving a reason. The list of available services that new technologies have increasingly offered for several years could be extended almost indefinitely. While the diverse offerings have changed all of our lives, the impact on cities is yet to be truly seen.

4.3 Creative Cities

Cities are creative places because a great deal of tacit and explicit knowledge is concentrated here. Explicit knowledge is retrievable and not bound to individuals. It can be disseminated via information channels and comprises facts that are available to any interested person. In contrast, tacit knowledge is bound to persons, contexts and places and cannot be transferred to another place. It can only be passed on in direct contact with other people. In cities with a division of labour and high density, there are many opportunities for face-to-face contact and the associated exchange of knowledge. One's own knowledge and the knowledge of others gives rise to knowledge advantages and synergy effects that can lead to innovations and ideas in many areas. On the basis of local networks, an urban milieu is created in which further knowledge is generated that is crucial for the success of a region (Brake 2012, pp. 25–27; Helbrecht 2005, p. 124; Kujath 2005, p. 25).

But why are some cities more successful than others? This is a question that politicians and academics, as well as citizens, ask time and again. There is a broad consensus that university locations have greater development potential than cities without higher education institutions. Moreover, it is considered proven that cities are not only good for immigrants, but immigrants are also good for cities. In the U.S., immigrants bring a wide variety of skills with them and want to move up the ladder no matter what. In cities, they meet entrepreneurs who have great interest in cheap labor and offer many easy jobs for unskilled workers. Businesses and immigrants alike benefit. Since immigrants have the opportunity to choose from a wide range of different jobs, they can find the job that suits them best. There are also many opportunities for further education. The chances for socio-economic advancement are therefore optimal in cities (Glaeser 2011a, pp. 78–81).

Recognized universities such as the East Coast universities Harvard and the Massachusetts Institute of Technology (M.I.T) near Boston, Princeton (Princeton, NJ), Yale (New Haven, CT), Brown (Providence, RI) and Duke (Durham, NC), the University of Chicago and Michigan University in Ann Arbor, MI, in the Midwest, and the California universities Stanford and Berkeley (both near San Francisco) as well as the California Institute of Technology in Pasadena are regarded as cadres and starting points for innovative spin-offs. Stanford University, located in Palo Alto, has influenced the development of an entire high valley south of San Francisco (◻ Fig. 4.8). Among others, William Hewlett and David Packard (founders of Hewlett Packard), Larry Page and Sergey Brim (founders of

4

☐ **Fig. 4.8**   Stanford University and view over Palo Alto

☐ **Fig. 4.9**   Headquarter of Cisco Systems in Silicon Valley

Google) and Sandy Lerner and Len Bosack (founders of Cisco Systems) studied here (▶ www.stanford.edu) (☐ Fig. 4.9). Because of the high concentration of firms in the computer and electronics industries, the name Silicon Valley came to be applied to the region as early as the 1980s. The city of San Jose also owes its rise to become one of the largest cities in the USA in just a few

decades to the positive impact of Stanford University. Boston has benefited from the region's well-educated population for much longer. In 1636, John Harvard, educated as a Protestant priest in Cambridge, England, bequeathed his library and lands to a Latin school founded shortly before, which was renamed Harvard University in 1639. Beginning in the mid-nineteenth century, in Boston and elsewhere in the United States, the various religious communities caused the number of colleges, which often became universities, to increase. The Universalists founded Tufts University in 1852, Boston College was founded by the Jesuits in 1863, and Boston University was founded by Methodists in 1871 (Glaeser 2011a, p. 234). Already in the early stages of industrialization, the US recognized the connection between academic research and economic progress. In 1862, at the initiative of President Lincoln, the U.S. government set aside land for the construction of universities that would specialize in innovations in agriculture and engineering. Many of the universities founded during this period, such as MIT, Cornell University in Ithaca, NY, the University of Berkeley in California, and the University of Michigan, are now among the most prestigious teaching and research institutions in the world.

There are a number of cities in the U.S. that seem to magically attract well-skilled workers, while other locations are virtually being abandoned by the highly skilled. On the winning side are Austin, Atlanta, Boston, Denver, Minneapolis, San Diego, San Francisco and Washington, D.C., as well as Raleigh and Durham (both NC). These cities have experienced a brain gain, which is reflected in positive economic development. A brain drain, i.e., a loss of mostly young people with a good or very good education, has been experienced by Baltimore, Buffalo, Cleveland, Detroit, Milwaukee, Miami, Newark, Pittsburgh, and St. Louis. Here incomes have dropped and the economy is facing many problems (Harden 2004). Cities now know they are in competition for the brightest minds in the country and are investing in developing excellent research facilities. However, attempts to copy the success of Silicon Valley and promote it through targeted infrastructure development and other measures have often failed. It seems that a certain innovative milieu, fierce competition, and social dynamics that cannot be artificially created are far more crucial to a region's success than good infrastructure (Bettencourt and West 2011, p. 52). New York Mayor Bloomberg (2011, p. 16) has attributed the tremendous innovation potential of Silicon Valley and Boston to the two regions' good universities. Nevertheless, he is proud of the fact that in 2010 New York reportedly received more venture capital than Boston for technology start-ups.

4.3.1 Creative Class

According to Richard Florida (2002, 2005), many cities lack members of the creative class. It is of little help if good universities educate highly qualified academics, but they leave the city after completing their studies, and if universities fail to commercialise their knowledge. Young and creative people want to live in tolerant cities with a very wide range of leisure activities and diversified cultural offerings. This includes a lively music scene, ethnic diversity and an openness to different ways of life. Members of the creative class work in technical professions, in the financial sector, in the media, as artists or in the leisure industry (◘ Fig. 4.10). They may be writers, university professors, analysts, or thought and opinion leaders in the broadest sense. They themselves do not consciously feel that they belong to a particular class, but they cultivate a work ethic that values creativity, individuality, and achievement. Creativity is not part of their job description, yet their creative ideas generate value (Florida 2002, p. 48). Economic progress is only possible in locations that have certain prerequisites such as technology, tolerance, and creativity: *"The key to understanding the new creativity and its effects on economic outcomes lies in what I call the 3 T's of economic development: technology, talent, and tolerance"* (Florida 2005, p. 37). The so-called *Gay Index* and the *Bohemian Index* represent an important indicator of the tolerance existing in a city. Homosexuals (gays) on the one hand prefer to live in tolerant cities and on the other hand contribute to the economic boom. In the 1990s, four of the top 10 regions in the *Gay Index* were identical to the top 10 regions for the highest growth in the high-tech sector. The *Bohemian Index* is an even more accurate indicator of a region's economic potential than the *Gay Index*. It takes into account the percentage of writers, designers, musicians, actors, dancers, painters and similar professions in a society, which was particularly high in the 1990s in San Francisco, Boston, Seattle and Los Angeles, but also in much smaller cities such as Boulder and Fort Collins, CO, in Sarasota, FL, in Santa Barbara, CA, and in Madison, WI. On the basis of the above-mentioned indicators and other key figures, some of which are very different, e.g. for coolness, professional sports or overall environmental quality, Florida has created a *creativity index* that is intended to provide information on the innovative capacity and success of a region in a similar way to a barometer (Florida 2002, pp. 237–254).

4.3.2 Criticism

Richard Florida, who now teaches and researches at the University of Toronto, markets his ideas at symposia around the world and advises cities and regions in search of the royal road to economic recovery and a better future. From a scientific point of view, however, his theses are highly controversial. One problem is that employees in the arts sector can hardly be clearly identified. Should all those who call themselves artists or who work in the entertainment sector in the broadest sense be taken into account, or only those who belong to high culture? Moreover, it is not the relative proportion of those employed in the creative industries that is relevant; far more telling is the absolute figure. According to Florida's analysis, relatively speaking, more people in Buffalo, NY are among the super-creatives than in New York City. Yet Buffalo has yet to stand out as a particularly creative city. In his books, Florida presents many statistics on the creative class and creative economy in individual cities, but fails to provide evidence of whether and how they actually stimulate the economy. Many of the cities that Florida has rated as particularly creative have actually performed worse than comparable cities. Albany, NY, and Dayton, OH, are among the midsized cities that Florida has rated as particularly creative. In fact, however, these cities' labor markets have under-

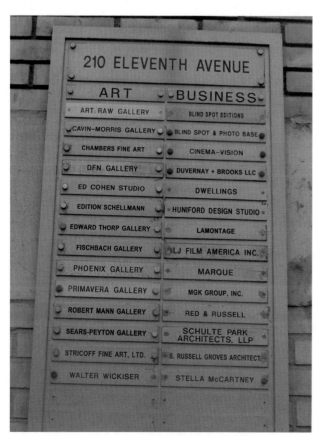

◘ **Fig. 4.10** Doorbell sign in Chelsea (Manhattan)

4

performed (Malanga 2004; Rushton 2009, pp. 164–166, 178). Economic growth only occurs when members of the creative class actually develop new ideas and implement them (Ross et al. 2009, pp. 71–72). Moreover, it is important to remember that artists, who according to Florida are particularly important to the future of a location, have not moved into run-down working-class neighborhoods in protest against the established, but because of the cheap rents (Zukin 2010, p. 22). In these neighbourhoods, the creatives have indeed initiated *gentrification*, but this is not measurable, but has led to qualitative changes. In- or trendy neighbourhoods have emerged that are distinct from the normal. As urban living environments have become more contradictory, ambivalent and pluralistic, they offer a good breeding ground for creativity. The way has been paved for a *"renewed appreciation of inner-city lifestyles for certain sections of the population"* (Gerhard 2012, p. 62).

It is also contradictory that Florida's model of urban and economic renaissance promotes *gentrification*, but at the same time he laments that as a consequence of this very *gentrification* and the associated higher rents many Bohemians leave the city again (Lees et al. 2008, p. XX). Kotkin and Siegel (2004, p. 57) also do not fully agree with Florida's thinking on the creative class. They believe that a mix of different common sense decisions determine the attractiveness and success of cities. These include innovative schools, good police and fire departments, and the ability to purchase affordable housing. Malanga (2004) goes further and speaks of the *"curse of the creative class"*. Florida's theses have been gratefully taken up by liberals or left-wing politicians in particular. This has led many cities to invest in the arts sector or expensive biking and walking trails in order to attract the creative class instead of giving the tax money to innovative companies. According to Piiparinen (2013), Florida has cleverly managed to first create a doomsday scenario and then show a way out of the impending crisis that no one can escape, as no one wants to be seen as uncool, intolerant or not creative.

4.4 Consumer and Tourist Cities

A number of US cities have grown up with tourism. Thousands or even tens of thousands of hotel rooms, a large number of restaurants, fast-food providers and a wide variety of entertainment facilities try to attract as many visitors as possible. This is true for Miami Beach, Las Vegas, or Waikiki, which is part of Honolulu in Hawaii ( Fig. 4.11). In addition, consumer cities have emerged in more recent times. With deindustrialization and the shift to a service economy, new elites have emerged who use cities differently than workers used

◘ Fig. 4.11 View over Waikiki (Hawaii)

industrial cities before them. Immigrants used to move to cities because of their large supply of industrial jobs. Today's highly educated service providers are highly educated and earn well, they are consumerist, and they are largely free to choose where they live. The elites have developed a cosmopolitan lifestyle that cities must satisfy. Under the influence of social change and global development trends, consumption has become more important than production. In a society with many singles and childless couples, the most important factors in finding a place to live are no longer good schools and churches, but recreational facilities. Cities like New York, Chicago, and San Francisco have responded to the changing consumer habits and lifestyles of residents. Consumers want organic food, cafes, weekly markets, galleries, and theaters; so these things are provided. Cities that change in response to the demand of their residents become consumer cities (Rushton 2009, p. 168; Zukin 2010, p. 27).

Only a few decades ago, growth and success were dependent on the creation of new industrial jobs, while in more recent times those cities that served the desires of the new elites were particularly successful (Fainstein and Judd 1999, p. 2). On a national and on a global scale, cities are competing with each other in ways previously unknown. Television channels such as CNN or MTV and the omnipresent movies from Hollywood create homogeneous consumption patterns worldwide. Needs are awakened that must be met in order to attract highly qualified residents and tourists. Soft location factors have become more important than hard ones. These can include a pleasant climate, a low crime rate and an associated sense of security, cleanliness and attractive public spaces, parks and jogging trails, as well as a wide range of cultural offerings and good schools and universities. In addition, a wide range of entertainment and leisure facilities must be provided to keep the new

◻ Fig. 4.12 Navy Pier in Chicago

◻ Fig. 4.13 Street café on Chicago's North Side with Rolls Royce

elites happy (◻ Fig. 4.12). Special events and icons promote global marketing (Clark et al. 2002, pp. 494–499). As the expansion of leisure facilities is a top priority in many cities, Clark (2004, p. 8) refers to them as *"entertainment machines"*. The term *"machine"* is meant to illustrate that entertainment is no longer a private matter, but has advanced to a commodity that cities must produce. It has become the most important instrument of urban development policy. In the past, municipalities encouraged manufacturing to create jobs; today, low-skilled immigrants work in the leisure industry (Clark et al. 2002, p. 499). Cities respond to consumer needs by sponsoring advertising campaigns and major events and by entering into public-private partnerships with private investors to jointly build hotels, convention centers, and shopping centers (Fainstein and Judd 1999, p. 2).

Politicians never tire of touting the virtues of inner-city *entertainment districts to* attract capital and visitors. The cities are to become as clean as the suburbs. This development began in the 1970s, when the waterfronts once used by industry and commerce were converted into *festival markets* (Zukin 2010, p. 5). It is the task of cities to create the institutional conditions for the transformation from industrial to consumer cities and to provide important public goods such as museums, parks or sports facilities. Private investors complement the public facilities with cafés, art galleries, shops or specific architecture, which should give the cities a distinctive look. The cultural and leisure offering, together with a corresponding income, without which the diverse offering cannot be used, determines the value of a city (◻ Fig. 4.13). The biggest winners of this development are the affluent, while the destitute are largely excluded from enjoying the diverse opportunities (Clark et al. 2002, pp. 494–497). Upon closer examination, however, even the less affluent benefit, as there are inexpensive restaurants and even free amusements in some cases, such

as Millennium Park in Chicago or the new Citygarden with its many works of art in the heart of St. Louis. Visiting many restaurants, bars, nightclubs and theatres is expensive, but they are important for revitalising inner cities at late hours (Glaeser 2011a, p. 11).

Cities have become a stage: Seeing and being seen is more important than ever. Following the demands of postmodern society, cities are used differently than before. The new urban elites like to frequent the numerous restaurants in the cities and rarely cook for themselves. Shopping is no longer for sustenance, but has become a pleasant pastime. Retailing or shopping and entertainment merge into *retailtainment* or *shoppertainment* (Hahn 2001, p. 20). Inner-city retailers cater to this trend with large flagship stores and shopping centers that offer many attractions. Shopping has become an important leisure activity for well-off urbanites as well as for the many suburban visitors and tourists. New York has particularly benefited from this development. The city has become an international shopping mecca. From 1998 to 2007, the number of those working in clothing or accessories stores in the country's largest city increased by half (Glaeser 2011a, p. 126).

4.4.1 Congress Tourism

The inner-city hotels, shops and restaurants profit from congress participants with purchasing power. In many inner cities, the numerous hotel towers, each with hundreds of beds, are only fully booked during major trade fairs and can then charge correspondingly high prices. Nationally and internationally well-connected airports such as those in Atlanta and Chicago are further important prerequisites for becoming a major convention location. But conventioneers want to meet and enjoy themselves in attractive locations or escape the cold

◻ Table 4.1 Top 10 convention centers in the USA 2012

Convention center	Rank[a]	
	North American Business Review[b]	Cvent Event Management[c]
Orange County Convention Center, Orlando	1	1
McCormick Place, Chicago	2	5
Las Vegas Convention Center	3	3
Walter E. Washington Convention Center, Washington, D.C.	4	2
Dallas Convention Center	5	9
Georgia World Congress Center, Atlanta	6	8
Phoenix Convention Center	7	7
Moscone Center, San Francisco	8	
Anaheim Convention Center	9	
San Diego Convention Center	10	6
Miami Beach Convention Center		4
New Orleans Ernest N. Morial Convention Center		10

[a]The criteria for ranking are unknown
[b]► www.businessreviewusa.com
[c]► www.cvent.com

weather of their hometowns in the winter. Since Orlando has good transport access as well as good weather and also has several attractive theme parks with Seaworld, Legoland, Universal Film Studios Florida and Epcot, it is now the most important trade show location in the country. In order to attract as many large trade shows and conventions as possible, a sort of arms race is taking place between cities and convention centers are continually expanding. There are different statements in various rankings about which convention centers are leading (◻ Table 4.1). Since all the data on visitor numbers differ widely, they are not mentioned here.

4.4.2 Attractions

Consumer cities address their own population, residents of suburban areas with purchasing power and tourists. International visitors are only attracted by well-known sights, attractions or events. With a few exceptions, such as the skyscrapers in New York or Chicago or the Golden Gate Bridge in San Francisco, there are few structures in American cities that can be marketed worldwide. New York, with its numerous landmarks, museums and Broadway shows, has by far the greatest appeal. In 2010, 8.4 million foreigners visited the city, followed by Los Angeles and Miami with 3.34 million and 3.1 million foreign tourists, respectively. Faceless cities try to compensate by buildings constructed by international architects or by historicizing buildings and structures, as they have been since the 1970s when *festival markets* such as Pier 39 in New York, Fisherman's Wharf in San Francisco, Quincy Market in Boston, or the Inner Harbor in Baltimore were built. Other cities have designated *historic districts* or *art districts* with old train stations or warehouses converted into multifunctional buildings with shops, restaurants and apartments. But what works in one city may be a failure in another. While *festival markets* in Boston and Baltimore have long been considered a great success, similar projects in Toledo, OH, Richmond, VA, and St. Louis, MI, have never lived up to expectations (Fainstein and Gladstone 1999, pp. 22–23; Fainstein and Judd 1999, p. 9). In *historic districts*, cities seek to evoke the past, albeit often with embellished or even false stylistic elements. Themed worlds, first developed at Disneyland and Disneyworld and in the construction of casinos, often lack any reference to reality. The result is the non-places described by Marc Augé (2011, p. 97), which have become interchangeable, similar to the huge check-in halls of airports, and are valorized by visitors on the basis of instruction manuals. In 1976, New Jersey became the second US state after Nevada to allow casino construction, but limited it to Atlantic City. Many financially weak cities such as Detroit and New Orleans have since followed the example of the once popular seaside resort. The casinos are intended not only to make cities more attractive, but also to fill empty municipal coffers at the same time (Leiper 1989). The city with the most casinos is still Las Vegas, which now offers much more than just gambling. The casino sites that have sprung up across the US since the late 1970s aim to copy Las Vegas, but are ultimately just cheap knock-offs of the venues in the world-famous tourist city. Especially in the Detroit casinos, the lower class dominates, hoping to escape misery with a win, but in fact being the big loser (author's observation).

All major American cities pursue the same goals. With the same elements over and over again, such as *festival markets*, *historic* and *art districts* and themed worlds, staged urban spaces are created that ultimately serve to stimulate consumption. Since modern man always demands something new, all attractions have to be constantly adapted to the zeitgeist at high cost. Perhaps the biggest problem is that cities will eventually all look the same and thus be interchangeable. Furthermore, if the desires of tourists take center stage, cities themselves may become theme parks and at some point locals will no longer feel comfortable in their own city (Fainstein and Judd 1999, pp. 12–13; Holcomb 1999, p. 69; Sassen and Roost 1999).

4.5 Downtowns

The city centers are the region's shop window. Here, the process of urban decay was reflected early on, but revitalization also often began in central city areas. In the United States, the city center is colloquially referred to as downtown. The term originated in the nineteenth century in New York, where in Manhattan, as development expanded northward, people increasingly spoke of downtown, midtown, and uptown to distinguish the three subareas. Soon other cities were also referring to their historic core as downtown. In some cities, however, other expressions for the central areas of the city prevailed, such as *Loop* in Chicago or *center city* in Philadelphia (Caves 2005, pp. 130–131). All downtowns are characterized by a number of common features. They are clearly set apart from their surroundings in terms of construction and function, and they are almost always the oldest part of the city. The first buildings were erected here, and city halls and large parts of the public administration are still located in the downtowns today. The building density is high and the skyscrapers can be seen from a distance. In the nineteenth century, the railway lines led to the city centes in a star shape, which is also true today for the public transport lines and the most important roads. Most downtowns still have the old train station, although it is rarely served by passenger trains today. There is also retail, banks, office buildings, museums, hotels, parking garages, recreational facilities and, in some cases, parks and apartment buildings. In the larger downtowns, the various uses are concentrated in their own neighborhoods. Office and retail rents are highest in this location. The adjacent *zone of transition* is structurally and functionally different from the downtown, with most buildings having only one or a few stories and serving as warehouses, auto repair shops, or other small businesses. Since the 1950s, ring roads have also been constructed in many cities to separate central city areas from surrounding neighborhoods (Caves 2005, pp. 130–131; Sohmer and Lang 2003, p. 64). The term *central business district* (CBD) is sometimes used interchangeably with downtown or to refer only to the central business areas of downtown. Downtowns as well as CBDs cannot be precisely delineated because they are not included in the census as a statistical unit. In addition, census tracts are continually adjusted to reflect development. Downtowns may shrink or increase in area over time. Therefore, it is not easy to accurately track and represent their development. What is certain, however, is that downtowns were long characterized by processes of decay, which in recent times have been replaced by an upgrading of at least some of the cities.

4.5.1 Loss of Importance and Decline

Although the first innovations in transport and in the construction of new houses (skyscrapers, elevators) began to take hold in the late nineteenth century, the patterns of urban land use initially changed little. However, by the 1920s, when the automobile became the most desirable object for Americans, the decline of downtowns was almost unstoppable, and retailers watched with alarm as more and more stores opened along arterial streets and competition increased (Muller 2010, pp. 307–310). At the same time, the image of towns was rapidly deteriorating. In the first two decades of the nineteenth century, hundreds of thousands of blacks, almost all of whom were uneducated and unaccustomed to urban life, had migrated to the industrial cities of the North in search of work. The blacks invaded inner-city residential neighborhoods, and soon tensions arose with the resident white population that not infrequently ended violently. Well-known are the riots and fights, some of which resulted in many deaths, in St. Louis in 1917 and shortly thereafter in Chicago and Washington, D.C., whites fled their homes in central city areas, where blacks subsequently moved in. Large apartments were often subdivided into multiple units, and overcrowding occurred due to the lack of housing for the new arrivals. What had once been middle- and upper-class housing degraded, and the first slums began to form. In the early 1940s, a study by the Urban Land Institute had found major blight and falling land prices in seven selected downtowns, prompting a Central Business District Council in 1946 to address the problems of the downtowns. Several cities decided to conduct cleanup and beautification campaigns, promote retail, or build high-rise apartment buildings, some of them high-priced, to halt further decay and attract higher-income residents. Some cities also planned to demolish inner-city slums (McKelvey 1968, pp. 17–22, 134–135). However, these plans were rarely implemented at the time because of a lack of financial resources and legal ability to enforce them. At the same time, suburbanization and the decline in the importance of downtowns accelerated. Downtowns were failing to attract investment, retail was retreating, and new office buildings were springing up on the periphery. The signs of decay could no longer be ignored, and real estate and land prices plummeted. At the same time, as the image of downtowns continued to deteriorate, they were con-

4

sidered dangerous, and all needs could be met in the surrounding areas of the cities, there was little reason to visit the inner cities (Kromer 2010, p. 51). By the mid-twentieth century, it seemed as if downtowns had entered a downward spiral from which there was no escape. It is therefore all the more surprising that at least some of the cities today have extremely attractive downtowns with very high land prices, which have become popular residential locations and magnets for tourists.

4.5.2 Population Development and Residents

In many downtowns, the number of residents has increased significantly since the 1980s, and this process has even accelerated in recent times, while other downtowns have stagnated or continue to lose population. Since the boundaries of downtowns are not precisely defined, it is necessary to rely on case studies that nevertheless reveal clear trends.

Using census data for the years 1990 and 2000, Sohmer and Lang (2003) examined the development of 24 downtowns of predominantly very large cities such as Chicago, Houston, Atlanta, and Los Angeles. The boundaries of these downtowns had previously been defined as part of a project based at the University of Pennsylvania. The area of each downtown varied widely: San Antonio had the largest downtown at 14.3 km^2, while those of Cincinnati, Lexington, and Norfolk occupied an area of only 2.1 km^2. Over the study period, only the downtowns of Charlotte, San Antonio, Phoenix, Lexington, St. Louis, and Cincinnati had lost residents, while all others had gains, although there was no compelling correlation between a downtown's population gain or loss and the overall city's population growth. Phoenix experienced a 34% population gain from 1990 to 2000, while downtown lost 9% of its residents. On the other hand, the downtowns of Cincinnati and St. Louis experienced increases in population while the cities as a whole lost population (Sohmer and Lang 2003, pp. 65–70). A similarly mixed trend was found for the decade from 2000 to 2010. In New York and Los Angeles, population grew in the city as a whole and in the downtown area. St. Louis lost about 36,000 residents, but gained 7000 people downtown. In Chicago, the overall city's population losses were far greater, but so were the gains of the *Loop* (Kotkin and Cox 2011).

An interesting approach was taken by the U.S. Census Bureau (2012, pp. 25–28), which analyzed population trends within a two-mile (3.2 km) radius of the city halls of the largest city in a metropolitan area for the period from 2000 to 2010 (◻ Table 4.2). MAs with a population of at least five million experienced the largest per-

centage increases around city halls, consistently in the double digits here. For MAs with a population of 2.5–5 million, a positive trend was also observed in the central areas, but this was much smaller in relative terms, while in the even smaller MAs a loss of population was particularly frequently observed near the town halls. However, even these data cannot be generalized, as they differ greatly in individual cities when examined more closely. By far the largest gain was seen in downtown Chicago, at just over 36%, where more than 180,000 people now live near City Hall; in New York, Philadelphia, and San Francisco, the number is actually much higher; though here the increase data was smaller due to the high baseline. The loss of residents in downtown New Orleans can be attributed to Hurricane Katrina in 2005. It is surprising, however, that Baltimore has also lost more than 10% of its population around City Hall, even though many apartment buildings have been built in the Inner Harbor area in recent decades (see below).

Who are the new urbanites? Are they predominantly young, well-educated people (*young urban professionals*) who do not yet have children, or couples whose offspring have left home (*empty nesters*). For several decades, downtowns attracted mainly financially weak immigrants because of the large supply of cheap housing. This has increasingly changed in recent decades, as a survey conducted in 44 downtowns shows. In the center of 20 of these cities, such as Dallas, Miami or midtown Manhattan, there is at least one census tract with a higher median income than in any other location in the region. These people could live anywhere, but have voluntarily chosen to live downtown. In addition, the percentage of young people ages 25–30 who have at least a college degree has increased significantly from only 13% in 1970 to 44% at the turn of the millennium (Birch 2005). Some of the inner cities have also benefited from the influx of older people. After children have moved out, many *empty nesters have* realized that living in the city offers advantages such as proximity to work, restaurants, and cultural amenities and apartments without stairs or large lawns to mow regularly (Bridges 2011, p. 97). The number of apartment owners in downtowns more than doubled from 1970 to 2000 to about 22%, although there were large regional differences. While 41% in Chicago owned the property they occupied in 2000, this was true of only 1% of Cincinnati's downtown residents. The ethnic composition of the inner city population is interesting. From 1980 to 2000, the combined share of black and white residents had fallen from 81% to 73%; however, the share of whites had risen again in the 1990s (Birch 2005).

Downtowns have largely lost their former function as the dominant office location in the region. Although the major US cities regularly report an increase in jobs

□ Table 4.2 Population development within a radius of 3.2 km around the city halls of the largest cities of selected metropolitan areas

Metropolitan area	Bevölk. 2010 (in millions)	Population within a 3.2 km radius of the town hall		Change 2000 until 2010	
		2000	2010	Number	%
Largest gains					
Chicago-Joliet-Naperville, IL-IN-WI	9.5	133,426	181,714	48,288	36.2
New York-Northern New Jersey-Long Island, NY-NJ-PA	18.9	400,355	437,777	37,422	9.3
Philadelphia-Camden-Wilmington, PA-NJ-DE-MD	6.0	214,760	235,529	20,769	9.7
San Francisco-Oakland-Fremont, CA	4.3	336,092	355,804	19,712	5.9
Washington-Arlington-Alexandria	5.6	137,064	156,566	19,502	14.2
Biggest losses					
New Orleans Metairie Maven, LA	1.2	116,193	80,880	−35,313	−30.4
Baltimore-Towson, MD	2.7	165,970	155,776	−10,194	−6.1
Dayton, OH	0.8	51,218	41,053	−10,165	−19.8
Toledo, OH	0.7	65,857	55,739	−10,118	−15.4
Saginaw-Saginaw Township North, MI	0.2	49,678	40,004	−9674	−19.5

Source: U.S. Census Bureau (2012, p. 28)

in downtowns in the press, these figures do not always stand up to scrutiny. Moreover, a not insignificant portion are jobs in menial and low-paying services, such as hotels, restaurants, or cleaning services. What is clear, however, is that an increasingly small proportion of a region's jobs are in city centers, as most new jobs have been created elsewhere in recent decades. In addition, jobs have moved out of city centers to other parts of the city or to suburban areas. Downtown Los Angeles lost about 200,000 jobs from 1995 to 2005, and Manhattan had nearly 42,000 fewer jobs in 2009 than in 2000. At the same time, the number of jobs in the other New York City boroughs has increased. A similar trend could be observed in Chicago (Kotkin and Cox 2011). A study of regional job trends in downtown, remaining core, and suburban areas in the 100 most populous MAs from 2000 to 2010 showed that only nine urban centers had not seen a decline in the share of jobs. Moreover, only in downtown Washington, D.C., had the absolute number of jobs increased (Kneebone 2013, pp. 5–6).

4.5.3 Revitalisation

The conditions for successful revitalization varied from city to city. This applied to the proportion of poor housing stock as well as the degree of slumming of former middle- and upper-class residential areas, the ethnic composition of the downtown population, the extent of the civil rights movement riots in the 1950s and 1960s and the associated degree of destruction and image loss, and the number of existing attractive buildings that were worth preserving. Some cities, such as San Jose or Fort Worth, have never had a significant downtown because they were laid out very late and then grew rapidly. Fort Worth even built an artificial downtown with a pedestrian mall and buildings from different eras to make up for this deficit (Newman and Thornley 2005, p. 58). Philadelphia was fortunate to have several attractive downtown residential sites that had never been completely abandoned by their residents. The elegant residences in Rittenhouse Square were always coveted by

4

the wealthy. Society Hill was able to benefit from a late 1950s revitalization program that had allowed for the rehabilitation of older homes and the filling of vacant lots. Residents of the aforementioned downtown neighborhoods consistently guaranteed minimum use of the various downtown amenities and a degree of vibrancy (Kromer 2010, pp. 50–51).

There has been a wide range of options available for downtown revitalization that have been used very differently. For example, there are cities with comprehensive land use plans and others that have, at best, loose zoning requirements. The same is true for the preservation of historic buildings. Also, the *urban-renewal* laws of the 1950s, the *CDBG Program* established in 1974, and the *American Recovery and Reinvestment Act of* 2009 have given municipalities a great deal of leeway in implementing policies. In addition, public-private partnership investors were not equally available in all cities and could have helped finance each project. In addition, the commitment and assertiveness of municipal politicians, and in particular mayors, varied greatly. Ultimately, the individual cities have implemented very different measures to revitalize their downtowns, which have been more or less successful (Fig. 4.14).

Beginning in the 1950s, federal *urban renewal* and highway expansion laws made possible the elimination of inner-city or near-city degraded residential areas that often wrapped like a ring around city centers. The construction of ring highways was financially supported and solved two problems at the same time: on the one hand, many of the unattractive residential areas disappeared and, on the other hand, the downtowns were better connected to the interurban road network. Funding from the *urban-renewal* program was explicitly granted only if a deteriorated and degenerated neighborhood was to be eliminated, but cities were given wide latitude in designating the sites. In addition, the brownfields created

by the demolition were to be sold to private investors to build new housing on the same site. Even when the cities actually tried to sell them, they often failed to find developers, because no one wanted to invest in these locations in the 1950s and 1960s. Particularly problematic was that new affordable housing was rarely built in other locations for the displaced population. Shopkeepers and small businesses suffered similarly from the demolitions. The sufferers of the clear-cutting were clearly the poor and members of ethnic minorities (Frieden and Sagalyn 1989, pp. 22–37). It is estimated that some 1.5 million people were displaced by the demolition of their homes. Increasingly, fairly arbitrary designation or expansion of existing redevelopment areas occurred in order to preserve large tracts of land for promising new development or because demolition was subsidized by public funds. An unattractive patchwork of open spaces and still used or unused buildings of different eras emerged in the downtowns (Schneider-Sliwa 1999, p. 50; Schneider-Sliwa 2005, p. 163).

The cities adopted bonus programs relatively early on in order to attract potential investors by means of more intensive use of the properties. At the same time, attractive recreational spaces were to be created for residents and visitors to the downtowns. As in many other cases, New York played a pioneering role. According to the 1916 building code, it was possible to completely build over a lot. In order to guarantee a certain amount of light at street level, the individual floors were increasingly stepped upwards, which is why many buildings resembled wedding cakes. In the course of time, however, it was not uncommon for special permits to be granted for the construction of additional storeys. By the end of the 1950s, there was growing unease about this practice, for it had to be feared that the very compact and tall buildings would soon prevent any light from entering at street level. In 1961, therefore, the New York City Building Code was amended, setting no maximum building height, only the permitted FAR (*floor area ratio*) at 15. A lot that was built on for 100% building height, but only the allowed FAR, could have a maximum of 15 stories. However, if only half of the lot was built on, a maximum of 30 stories was allowed, and if only one-quarter of the lot was built on, a maximum of 60 stories was allowed (Fig. 4.15b). In addition, other incentives were created to increase the attractiveness of the city. If a publicly accessible plaza was created on the undeveloped parts of the site, further additional floors were allowed at a ratio of 1:10. The maximum FAR thus quickly reached 18. As the program was better received than had been suspected, the first drawbacks soon became apparent. On the one hand, while many small public spaces were being created in Manhattan, on the other hand, the overall building mass was increasing at an alarming rate. In addition, many of the new

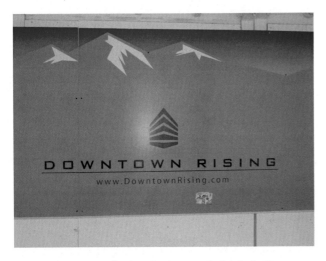

 Fig. 4.14 Motto of urban development in Salt Lake City

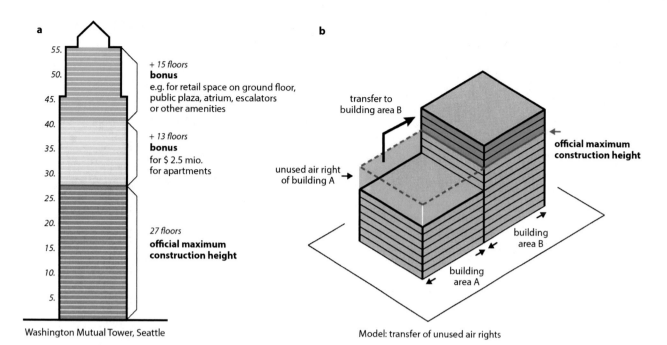

a

55.
50.
45.
40.
35.
30.
25.
20.
15.
10.
5.

+ 15 floors
bonus
e.g. for retail space on ground floor,
public plaza, atrium, escalators
or other amenities

+ 13 floors
bonus
for $ 2.5 mio.
for apartments

27 floors
**official maximum
construction height**

Washington Mutual Tower, Seattle

b

transfer to
building area B

unused air right
of building A →

**official maximum
construction height**

building
area B

building
area A

Model: transfer of unused air rights

Fig. 4.15 **a** Washington Mutual Tower. (Adapted from Cullingworth and Caves 2008, p. 113); **b** Air rights transfer and bonus floors. (Adapted from New York City, Department of City Planning 2013; keyword: *air rights*)

open spaces were extremely unattractive. Some were even unadorned concrete surfaces without any seating or shade. In 1975, therefore, design regulations were issued for the new public squares, and while not all investors complied, more interesting squares were now created overall. At the same time, however, a new practice crept in. Since investors understandably wanted to make as much profit as possible, they were interested in constructing very tall buildings. As a result, their building applications included more and more publicly accessible amenities such as conservatories or connecting walkways to the adjacent building. Since the city of New York was nearly broke, it readily agreed, because better utilization of land meant higher tax revenues. In the end, the maximum building height was negotiated individually for each building and, contrary to the 1961 building code, the city grew almost unrestrained into the sky and the incidence of light became more and more limited. This misguided development was not corrected until 1982. In individual subdivisions of Manhattan, the maximum FAR was now lowered, but it remained a controversial instrument. In addition, in New York, as in many other cities, so-called *air rights* can be transferred. If the owner does not exhaust the maximum FAR when building on a property, he can transfer the unused floor area to a neighbor, who is allowed to build correspondingly higher. It is particularly beneficial if there is a church or other protected structure on the neighboring property, because the unused FAR of these buildings can also be transferred (City of New York Planning Department 2013).

Seattle has a similar bonus program to New York, but also allows additional floors to be "*bought*" through donations for certain projects. If you make consistent use of all the possibilities in Seattle, the maximum height of inner-city buildings can be roughly doubled. In 1988, the Washington Mutual Tower opened as the second tallest building with a height of 235 m and 55 floors, 28 of which were due to the bonus program (Fig. 4.15a). 13 of the additional floors had been approved because the investor had donated $2.5 million for the construction of apartments. The increasing height growth has been criticized by many citizens. However, it became apparent that once incentives were introduced, they were difficult to reverse. It was not until 2001 that the often criticized program was successfully cut back (Cullingworth and Caves 2008, p. 113).

4.5.4 Megastructures

The new brownfields that had been created as part of the *urban-renewal* programs made room for new projects, often in line with the separation of functions proclaimed by the Swiss architect Le Corbusier. In the downtowns, huge complexes of buildings surrounded by open space, green space, and multilane streets were built to allow traffic to flow unimpeded. In the 1950s, the demolition of a railroad tunnel in the center of Philadelphia created a brownfield site of almost 9 ha on which Penn Center was built. The complex of several office buildings was connected by underground walkways to shops, restau-

4

rants and a subway station. At the time, Penn Center was considered one of the most successful projects for the revitalization of an American inner city. However, it did not make a lasting contribution to revitalization because it focused exclusively on its own interior (Kromer 2010, p. 52). In other cities, large-scale administrative centers with city hall and other government offices were realized with public funds. In Boston, a new Government Center with unadorned concrete buildings in the style of the time was built between 1963 and 1968 on the site of the former Scollay Square, which, with cheap bars and theaters, was in the eyes of many citizens an undesirable entertainment district on the edge of downtown (Kennedy 1992, pp. 176–181) (◘ Fig. 4.16). Older, positive mentions of the complex are hard to follow today, "... *the Government Center Plaza offers a sweeping open space dominated by the most spectacular new city hall in America ...*" (Gapp et al. 1981, p. 115), for City Hall looks extremely unattractive, and the huge central shadeless plaza is too hot in summer and too cold and windy in winter to linger. The Government District of Dallas and its central plaza, completed in 1978, have a similarly repellent effect, even though a sculpture by Henry Moore has been erected here and a water basin created.

In other locations, oversized building complexes were constructed that made no reference to their surroundings. Particularly off-putting are the Westin Bonaventure Hotel in downtown Los Angeles and the Renaissance Center in downtown Detroit (◘ Fig. 4.17), which are very similar and were both built by architect John Portman in the mid-1970s. Both buildings each consist of five separate circular, glass-fronted towers connected by a multi-story central atrium. The Westin Bonaventure has 35 floors and is the largest hotel in the city, with 1354 rooms. The Renaissance Center has as many as 72 floors and houses the headquarters of General Motors and a Marriott hotel. The numerous shops in the Renaissance

◘ **Fig. 4.16**   Government Center in Boston

◘ **Fig. 4.17**   Renaissance Center in Detroit

Center atrium have never been well received by customers, and the number of vacancies is high. The atriums of Portman's two buildings offer no views of the surrounding area and, with their blank concrete walls, look unattractive. Therefore, as part of a renovation of the Renaissance Center in 2004, a five-story glazed atrium opening onto the Detroit River was added. This conservatory is very popular with visitors and employees. Both buildings were celebrated as achievements of postmodernism on the one hand, but on the other hand they were criticized very early on because they seem like fortresses whose entrances are controlled by cameras and security guards. Renaissance Center employees live almost exclusively in suburban areas, driving to the complexes' parking garages, which they never leave, and back to the suburbs in the evening. "... *The message is clear. Afraid of Detroit? Come in and be safe.*" (Whyte 1988, p. 207).

The Renaissance Center is additionally connected to the neighboring buildings with glazed pedestrian bridges (*skyways*), which ensure that the street level and thus the (dangerous) public space really never has to be entered. Other cities even created entire *skyway systems* in the 1960s–1980s, connecting all the major office buildings, hotels, and shopping malls and parking garages in downtowns. The twin cities of St. Paul and Minneapolis in Minnesota have the most extensive systems of this kind. In downtown Minneapolis, some 70 city blocks can now be reached via *skyways*, which can be especially convenient during the very cold winters in this region. At the same time, the perfect car-friendly city has been created, as traffic can flow undisturbed under the *skyways*. Since downtowns are oriented solely to the interior and lack attractive storefronts at street level, there is no reason for pedestrians not to use *skyway systems*. Entire downtowns have become one giant megastructure, extremely hostile to the city as a public space (Hahn 1992, pp. 133–145; Whyte 1988, pp. 206–221).

Until the 1970s, the suburban area was the traditional location of shopping centers, which were increasingly built in downtowns in the following decades. The reasons why the operating companies suddenly invested in the inner cities are not entirely clear. Presumably, the good locations in the suburban area were spoken for in many regions and the investors saw the first signs of an upswing in the downtowns and sensed a future business. For the investors, the construction of these shopping centers represented a great risk due to the initially still very desolate condition of the downtowns and a lack of experience. Another disadvantage was that the land was very expensive and the construction costs were high, as the buildings had to be built high up. In addition, it was impossible to predict whether customers would accept the shopping centers, especially as there was no free parking. In many cases, the shopping temples were integrated into multifunctional buildings. The 74-story Water Tower Place on North Michigan Avenue in Chicago, which opened in 1975, is home to one of the first inner-city shopping centers in the USA. North Michigan Avenue connects the *Loop* with the affluent northern neighborhoods. The shopping center occupies the lower eight floors of the multifunctional building, providing space for about 100 smaller retailers and restaurants and two large-scale vendors. Office space can be found on the ninth floor, with a luxury hotel above with 431 rooms on 22 floors and 240 luxury apartments on the top 40 floors. While the interior of Water Tower Place is very elaborately designed with a lot of glass and chrome and still meets today's standards, the exterior is not very successful. At least some small shop windows were integrated in the ground floor, but apart from that, the facade is completely unadorned without any kind of loosening elements and creates a repellent impression. If one enters the building through the main entrance, one immediately sees security guards who unmistakably suggest that one is in a safe room. Contrary to all predictions, sales were high right from the start, and Water Tower Place even acted as the initial spark for the expansion of North Michigan Avenue. Soon other shopping centers opened here and international designers settled on North Michigan Avenue, which today is undoubtedly the most attractive shopping mile in the USA after Fifth Avenue in New York. In the meantime, shopping centers such as Pacific Center in San Francisco, Prudential Center in Boston, Manhattan Mall in New York or Riverwalk Market Place in New Orleans have been built in almost all downtowns of the major cities, but not all of them have been as successful as Water Tower Place in Chicago and some have had little positive impact on the surrounding area (Hahn 2002, pp. 58–60; Whyte 1988, p. 2008).

4.5.5 Paradigm Shift

It soon became clear that demolition and new construction could not achieve sustainable *urban renewal*, and as the destruction of mature neighborhoods increased, so did the number of opponents to this practice. Beginning in the 1970s, new guidelines for downtown revitalization slowly took hold. Moreover, since President Reagan cut federal funding in the 1980s, cities were largely left to solve their own problems. Many cities relied on tourism and conventions to revitalize downtowns and increase revenues. Downtowns were connected to suburban areas by rapid transit, and sports stadiums, *entertainment districts*, revitalized waterfronts, parks, and vibrant streets with outdoor dining were designed to make downtowns more attractive to locals and visitors. The construction of large-scale megastructures was replaced by small-scale measures that were intended to contribute to an increase in the quality of stay. At the same time, great importance was attached to cleaning up and decriminalisation and the associated image gain. The measures took effect in many cities and again attracted investors who built new hotels and apartment buildings in the downtowns. In addition, a larger number of high-rise buildings were built in the inner cities again for decades.

4.5.6 Sports Stadiums and *Entertainment Districts*

As early as the late nineteenth century, Americans' great enthusiasm for sport led to the construction of huge stadiums, initially open but soon covered, which were often built close to city centrer for reasons of accessibility. The older stadiums have since been replaced by more modern and larger stadiums, the form of which has been largely standardised as sport has become increasingly commercialised and is anything but identity-forming (Conzen 2010, pp. 435–436). In addition, new stadiums, not infrequently designed to contribute to the revitalization of nearby downtowns, were built. Since American sports fans are better behaved than European hooligans, the locations are actually well chosen because spectators can help revitalize downtowns during evening hours and weekends. In addition, the city centers are well connected to the supra-local road network and to public transport. However, the calculation has not always worked out. When the Red Wings hockey team threatened to leave Detroit in the 1970s, Mayor Young had Joe Louis Arena built on the edge of downtown for $57 million and leased it to the team at a bargain price. The Red Wings stayed in the city, but the cost to Detroit was enormous (Glaeser 2011a, p. 62). Indeed, the contests at Joe Louis Arena, Ford Field football stadium, and Comerica Park baseball stadium draw suburban residents by the thousands who normally avoid downtown Detroit. Visits to the surrounding restaurants and bars, however, are limited to a few hours (author's observation). In Kansas City, the construction of an expensive sports stadium was even a complete misinvestment. In 2007, a $276 million, 18,500-seat multi-purpose arena opened on the edge of downtown, its construction generously subsidized by taxpayer funds. Despite great efforts, no hockey or basketball team has been able to secure permanent use of the Sprint Center. The National Hockey League and National Basketball Association teams preferred smaller cities such as Memphis, Oklahoma City, or Raleigh where they would not have to compete with baseball or football for the favor of spectators, sponsors, and press (Schoenfeld 2009). To this day, only occasional concerts or similar events take place in the stadium.

In contrast, a clearly positive contribution to revitalization has come from the MCI Center, a multi-purpose arena that opened in 1997 with more than 18,000 seats between 6th and 7th St. NW in the heart of Washington, D.C., which has been called the Verizon Center since 2006. It hosts the home games of the Washington Capitals ice hockey team and the Washington Wizards basketball team, among many other events. Since the late 1990s, this once badly deteriorated part of downtown, which was also the site of Chinatown, has seen a lot of development. 7th St. NW now performs the function of an *entertainment district* with many restaurants and bars, as well as retail. A few years ago, Gallery Place, an *entertainment center* with 14 cinemas, restaurants and retail, was opened. The cinemas and the events that take place several times a week in the Verizon Center guarantee a permanent stream of visitors at a downtown location with direct subway access. A symbiosis of sports stadium and *entertainment district* has also been created on the outskirts of downtown Los Angeles. In 1999, the multi-functional Staples Arena was opened here and supplemented by a convention center on the opposite side of Figueroa Street. From 2005 onwards, L.A. Live was built in the immediate vicinity as part of a public-private partnership, with a Nokia Theatre, where the Grammys are awarded every year, two large luxury hotels and several restaurants grouped around the central Nokia Plaza. Oversized LED screens presenting large-scale advertisements are meant to be reminiscent of New York's Times Square. The latter is not really successful, as Nokia Plaza seems very artificial compared to the New York model, but at least a visitor magnet has been created in downtown Los Angeles with the Staples Arena and L.A. Live (observations by the author).

4.5.7 Clean Downtowns

In some cities, such as New York and Los Angeles, determined mayors have taken drastic measures to clean up and decriminalize inner cities. Mayor Giuliani's cleanup of New York's Times Square, which embodies a symbol of the decay and rise of the North American inner city, became particularly well known. A closer look, however, reveals that the world-famous square's resurgence had begun long before Giuliani took office. Giuliani, however, implemented a highly controversial *zero-tolerance strategy* that involved prosecuting even minor offenses such as drinking in public or aggressive panhandling in order to prevent major crimes from occurring in the first place (▶ www.krimilex.de, keyword: *zero tolerance*). In Times Square, the number of criminal offences was reduced from just under 4000 in 1993 to just under 800 in 2011 (Times Square Alliance 2012).

The rectangular street layout of Manhattan is broken up by Broadway, which runs diagonally from south to north in the style of an ancient Indian trail. 42nd St. forms the southern boundary of Times Square, which is the undisputed center of New York. With many restaurants, shops and entertainment venues, 42nd St. has long been considered the biggest magnet for New Yorkers and visitors to the city. Gradually, more and more of the ancestral establishments were replaced by cheap shops, seedy restaurants, and porn shops. At the same time, visitors changed, crime rose, and the image of Times Square and 42nd St. declined. Beginning in the 1950s, calls to "*clean it up*" demanded a cleanup of Times Square (Gratz and Mintz 1998, p. 69), which became a metaphor for solving the problems of the American city. Despite all efforts, the decay of Times Square could not be stopped for a long time: The porn shops and cheap bars and unwanted visitors remained. In 1981, the city of New York adopted a $1.6 billion plan to revitalize Times Square. At the same time, a new zoning plan allowed for greater building density. Public subsidies were to subsidize the construction of four new office towers, a large shopping center, and a 500-bed hotel. They also considered demolishing the Times Tower. However, the construction of the shopping center and the demolition of the skyscraper that gave the square its name were soon abandoned (Gratz and Mintz 1998, p. 70).

After years of preparation, the city of New York passed an ordinance in 1995 that barred porn shops, topless bars, and other shady establishments near churches, schools, and homes. By then, signs of recovery were evident. In the 1990s, Times Square also benefited from the boom in the media and telecommunications sectors and the positive performance of the economy, which supported the willingness of businesses to invest. In Times Square and along 42nd St., buildings in need of redevelopment were demolished and replaced with high-rises. At the same time, small-scale storefronts were displaced by large-scale retailers that benefited from rising tourist numbers. Investments by the Disney corporation gave 42nd St. a new look. The company bought and restored the Amsterdam Theater, which opened in 1903, for its own musical productions. Next door, Disney set up a large store selling the corporation's fan merchandise, which enjoyed great popularity, at least for a while. Disney stands for "*clean*" and "*safe*", and its professional management helped to fundamentally change 42nd St.'s image and attract the interest of other investors. Other major attractions included a huge Virgin Megastore and a sports bar owned by the ESPN television network and an MTV studio. For years, ABC has also broadcast the daily news show *Good Morning America* from Times Square. Commerce and the serious world of financial news merge in this square, where the technology exchange NASDAQ also maintains a glass television studio. The giant Disney store and Virgin Megastore are long gone. Disney had opened too many of these stores in the U.S., leading to oversaturation, while Virgin had to close due to declining demand for recorded music. In the meantime, however, Times Square has become such a major attraction that new tenants can be found immediately (Gratz and Mintz 1998, p. 71; Hahn 2001, author's observations).

Other cities, such as St. Louis, also do a lot to combat pollution, since only clean cities attract consumers ( Fig. 4.18). One smashed window that is not repaired is soon followed by others. Litter on the sidewalks tempts passersby to drop their trash instead of waiting for the nearest wastebasket. Littered streets are unattractive and create a sense of insecurity. A city that cannot maintain clean streets is unlikely to maintain law and order (Kelling and Wilson 1982). In many inner cities, such as St. Louis and Seattle, highly visible law enforcement officers maintain cleanliness. In Chicago, Mayor Richard M. Daley took up the fight against graffiti in 1993, claiming that it diminishes the quality of life and the value of homes and frightens people. Where one graffiti is not removed, more smears are soon to be found. In Chicago, a fleet of now 17 teams has been set up to remove graffiti using a mixture of baking soda and water under high pressure. If necessary, new paint is then applied. The free service is used more than 150,000 times each year. The campaign is supported by heavy fines for vandalism. In addition, the sale of spray paint is prohibited in Chicago. Many other cities such as New York, Los Angeles, Miami and San Antonio now also promptly remove any new graffiti, following Chicago's example (▶ www.cityofchicago.org). Another major problem is vacant houses, commercial establishments and shops that give an unkempt impression. Even the front gar-

◘ Fig. 4.18 Street cleaners in St. Louis

dens of still inhabited houses are often not maintained, but used as storage space for junk. In many places, abandoned cars with smashed windows or other signs of vandalism are parked in the public streets in large numbers. In 1998, the city of Philadelphia had 23,000 junk cars towed away, and 27,000 the following year. With cars being "*abandoned*" again every day, the problem was almost impossible to solve. By the turn of the millennium, there were an estimated 40,000 ownerless cars on Philadelphia streets and brownfields. In 2000, the new mayor, John Street, adopted a program previously announced during the campaign to more quickly remove these vehicles from public spaces. A hotline was established where citizens could report ownerless cars. All cars with an estimated value of less than $500 were to be removed. Vehicles with a higher residual value that gave the impression of being abandoned could be towed if the owners did not comply with the city's request for removal. If owners of abandoned vehicles were located, they could expect a fine. In the initial phase, 25 authorised companies towed away around 825 abandoned cars a day. Soon, however, word got around about the drastic measures and the owners drove their cars to the scrap yard themselves or parked them on their own properties (Kromer 2010, p. 110).

4

⬛ **Fig. 4.19** *Business improvement district* in Washington, DC

4.5.8 **Business Improvement Districts**

Many of the now successfully revitalised squares and shopping streets in the USA that were under pressure to devalue owe their success to their transformation into *business improvement districts* (BIDs) (⬛ Fig. 4.19). This is also true of New York's Times Square. In 1971, the world's first BID was established in Toronto, Canada, followed by the first U.S. BID in New Orleans in 1975 (Pütz 2008, p. 7). The legal basis for the formation of BIDs is regulated by the individual states and can therefore differ from one another. However, the formation always takes place through an association of all property owners and traders of a street, a square or a neighbourhood. All participants must pay into a common fund, from which certain services are financed, such as the maintenance of public safety by security guards or cameras, uniform signage and marketing, as well as the cleaning, lighting or greening of streets and squares. The services provided by the municipalities, which are actually responsible for some of these tasks, are thus improved. The public-private partnerships should lead to an increase in attractiveness and an increased number of visitors or customers (Armstrong et al. 2007, p. 1; Sorkin 2009, p. 83).

In New York, the Times Square Alliance BID was established in 1992, encompassing 123 blocks between 41st St. on the south and 53rd St. on the north, and Avenue of the Americas on the east and 8th Avenue on the west. The Alliance's annual budget is just under $12 million. The money not only pays for private security and cleaning of the plaza, but also funds countless events. The highlight every year is the New Year's Eve celebration with international stars. In addition, Times Square has been upgraded by many small measures. In 2009, the pedestrian areas were expanded and traffic was banned from some parts of the square. From 1992 to 2012, rents for retail space have increased tenfold to just over $20,000 per square foot, the value of all real estate has nearly tripled to $5.6 billion, and attendance at Broadway shows has increased by nearly 70% to 12.5 million (Times Square Alliance 2012). Times Square is now the biggest tourist draw in New York or even the U.S.. Because it is privately managed, the rules that are common in public spaces do not apply here. The square is now almost as safe as an amusement park or a shopping mall, which can also be criticized (Davis 2001, p. 5).

In New York, 67 BIDs are currently operating in all five boroughs, with more than $100 million available for a wide range of measures. Chicago has 44 BIDs, referred to here as *special service areas*, and Los Angeles currently has 31 BIDs. The number of BIDs in U.S. cities is now more than 1000 and rising. Of course, not all BIDs are as successful as the Times Square Alliance, but there is a consensus that they have made an important contribution to urban development. In New York, it has been shown that property prices in BIDs rise faster than in adjacent neighbourhoods or in comparable locations. This is particularly true for larger BIDs, where commercial properties were 15% more expensive than in other locations. The smaller BIDs have comparatively high administrative costs and have little funding for upgrading activities. In addition, the larger BIDs are more professionally managed than the smaller ones. However, residential property prices are hardly positively influenced by the formation of BIDs. It is even possible that the increased traffic volume and the associated burdens have a negative impact on the demand for housing in the BIDs (Armstrong et al. 2007).

4.5.9 **Waterfronts**

The North American city had long had no showside to the water. On the shores of the lakes and rivers, as well as the Pacific and Atlantic Oceans, ports had sprung up with the founding of cities, along with all the buildings that went with them, such as repair shops and warehouses, as well as cheap bars and seamen's amusement establishments. Since the 1950s, the historic harbours have suffered a loss of importance, as the ever larger ships could no longer be handled in the harbour basins and at the finger piers. The warehouses were also no longer needed, as goods were transported directly onwards or delivered in containers. At the same time, many industrial companies moved out of the port and into suburban areas. Unsightly and unused buildings remained on the sites of the former port facilities, which were at best replaced by large areas of wasteland. In the 1960s, the first urban planners and politicians recognised the development potential of the areas located on the edge of the city centers, which lent themselves to various new uses. Should new jobs be created here in the secondary or tertiary sector, should housing be built, or perhaps better, should leisure facilities be created? Several uses were always in competition with each other, and conflicts were inevitable. In addition, contaminated sites were problematic, as was the fact that the areas were often not abandoned as a whole, but gradually, and usually belonged to a larger number of owners (Hahn 1993).

All cities initially had to invest large sums in the old port locations in order to make them interesting for private investors. In Baltimore, between 1968 and 1983, about 45% of all investments were made by the city; between 1983 and 1987, this share was only about 5%. In San Francisco, Fisherman's Wharf, which is not directly adjacent to downtown but is strategically located at the terminus of the cogwheel railroad, has seen the conversion of old factory buildings and piers into shopping malls, restaurants and snack bars, and the creation of a marina since the late 1960s, creating a colorful fairground that is a major tourist attraction. In Boston and in Baltimore, the transformation of the old port facilities has occurred in conjunction with the redevelopment of the adjacent downtown areas (◘ Fig. 4.20). In both cities, shopping centers, museums, and aquariums, as well as residences and office buildings, have been developed in the area of the former port facilities since the 1960s, although in Boston, far more existing buildings such as Quincy Market, built in the 1820s, have been injected with new life than in Baltimore, where almost exclusively new buildings have been constructed. The opening of Quincy Market in 1976 on the edge of downtown Boston marked a turning point. Until then, downtowns' loss of prominence and decline seemed unstoppable. The mix of high-end fast food, owner-operated local retail,

◘ **Fig. 4.20** Inner Harbour of Baltimore

and large recreational offerings in historic market halls proved attractive to locals and tourists. Quincy Market was innovative and unique, and acted as an incubator for downtown revitalization (Gratz and Mintz 1998, pp. 48–49; Hoyle et al. 1988).

In the Port of Baltimore, which was once dominated by the steel producer Bethlehem Steel, handling and industry suffered a massive slump between 1962 and 1985, which initially particularly affected the port locations close to the city center. In the adjacent downtown, the loss of significance had begun in the 1940s and intensified in subsequent decades. After several attempts to revitalize downtown had failed, starting in the late 1950s some 700 buildings in the downtown area were demolished, though by no means all of them had fallen into disrepair or were unused, in order to build the multifunctional Charles Center here on 13 ha with several office buildings, retail and restaurants, which was given direct access to the subway. The Charles Center was connected to the surrounding blocks with pedestrian bridges. The redevelopment of the adjacent Inner Harbor had long been discussed, but was deemed too risky due to its poor image. In 1968, the first clearing and cleanup work finally took place here, and a redevelopment plan was adopted in 1971. The continuous promenade around the harbor basin was completed the following year, and construction of additional buildings or remodeling of existing structures soon followed. The Rouse Company constructed a *festival market* consisting of retail and food service buildings designed to resemble indoor markets. Furthermore, a large aquarium, the Maryland Science Center, and the Visionary Art Museum, which unlike the city's other museums does not exhibit established art, were built, and a football and baseball stadium were built west of the Inner Harbor. An old power plant was converted for bars, restaurants, several stages, and a large bookstore. Over the years, construction of a larger number of apartment buildings and hotels followed, not only in the Inner Harbor, but also in the adjacent near-shore areas of the Patapsco River. While hardly anyone strayed into the Inner Harbour at the beginning of the 1970s, around three decades later more than 22 million visitors were counted here annually, of which around 30% were tourists who often also spent one or two nights in the hotels (Pries 2009, pp. 91–103).

The revitalisation of the Inner Harbour has long been regarded as a major success, with ports around the world looking to it as a model for the conversion of their derelict areas. Remarkably, an increasing number of very high-end apartment buildings have been built in the waterfront areas. On the south shore, Ritz Carlton Residences have even been built with luxury apartments of up to around 500 m² of living space. However, the timing of the completion in 2008 at the height of the

4

real estate and financial crisis was conceivably bad. At the same time, a number of other apartment buildings were ready for occupancy but could not be sold or rented out. Although demand has since picked up, many apartments in Inner Harbour and on the banks of the Patapsco River were still for sale in early 2013 for an average price of just over $260,000. But just two or three blocks from the waterfront, prices are plummeting significantly, as there has been little new construction or redevelopment here. The revitalization of Baltimore's waterfront has not had a positive impact on the city as a whole, where apartments or single-family homes are averaging only $96,000 overall. This value is in stark contrast to Washington, D.C., only about 70 km to the south, where the comparable value is over $400,000 (▶ www.zillow.com). In some subdivisions, *gentrification* has also taken place. This is particularly true of the former working-class districts of Fells Point and Little Italy, where many of the small apartment buildings have been demolished or subjected to noble redevelopment, which has been associated with a major change in the population structure (Pries 2009, pp. 108–114).

The downtown Charles Center has already been remodeled several times, with only very hesitant success. The very plain Charles Center, built in the style of modernism with a lot of exposed concrete, was never really accepted outside office hours and the number of empty business premises was always high. As the Inner Harbour became more successful, the Charles Center's problems increased. It wasn't until a redesign of the exterior starting in 2009 that the mega-complex was able to increase its appeal. The Inner Harbour has long assumed the function of a huge *urban entertainment center*. However, the public is demanding more and more new attractions, and competition is fierce. In May 2011, it was therefore decided to have private investors build, among other things, a large Ferris wheel, 14 beach volleyball courts and a facility for bungee jumping. It seems that the Inner Harbour will be transformed into a funfair. It remains to be seen if and for how long the new attractions will be able to attract visitors. It is also possible that potential buyers of the luxury apartments will be discouraged from investing near the urban fairground (Hahn 2001; ▶ www.baltimoreredevelopment. org). Quincy Market in Boston has also long ceased to be the major visitor magnet it was in the first years after opening, as it has long ceased to stand out for its uniqueness as many cities have adopted the successful concept. Quincy Market has lost its identity and has become interchangeable (Gratz and Mintz 1998, pp. 48–49).

4.5.10   High-Rise Buildings

The term high-rise is relative and has changed over time. Surrounded by only four-storey buildings, a building with 15 floors looks like a high-rise; but if it is surrounded by 40-storey buildings, no one will pay any attention to it. Since the first high-rise buildings were built at the end of the nineteenth century, they have always been prestige projects designed to enhance a company's reputation. Even though they were mostly privately financed, they were always also the pride of cities that marketed themselves through their skylines, and skyscrapers became an American landmark early on. Because they seem to grow into the sky, they are known as skyscrapers. For a long time, New York and Chicago determined the competition for the tallest skyscraper (Lamster 2011, p. 80). Since the completion of the Empire State Building (381 m) in New York, first the Great Depression, then the Second World War and finally the loss of importance of inner cities had prevented the further construction of skyscrapers. It was not until 1972 that New York's World Trade Center (417 m) replaced the Empire State Building as the tallest building, but it was surpassed in 1974 by the Sears Tower in Chicago (442 m). On the West Coast, San Francisco's Transamerica Pyramid (260 m), a particularly striking building, was opened to the public in 1972 (CTBUH 2008). The skyscrapers erected in the early 1970s became famous because of their height, but they tended to be exceptions. It was not until the 1990s that a significantly higher number of skyscrapers were built again, which can be attributed to the increased appreciation of downtowns and the associated interest of investors in this location. However, skyscrapers were now increasingly being built on other continents and a skyline was soon no longer a unique selling point of the North American city, with the tallest buildings being built in Asia. In 1996, that record went to the Petronas Towers (452 m) in Kuala Lumpur. The U.S. was never able to claim the title again. In 1930, 99 of the world's 100 tallest buildings were in the U.S. and of these, 51 were in New York. By 2010, only 22 of the 100 tallest buildings were in the U.S. and as few as five were in New York. As of May 2013, the One World Trade Center in New York, which is still not quite completed, is the tallest building in the USA, expected to reach a height of 541 m (◘ Table 4.3, ◘ Fig. 4.21). Although the U.S. has clearly lost supremacy in this area, people still associate New York with the stunning Manhattan skyline. At the end of 2011, New York had 221 buildings taller than 150 m; 9% of all buildings of this height in the world or 34% of the U.S.. Only in Hong Kong can more buildings of this scale be found (CTBUH 2008, p. 41; CTBUH 2011, pp. 54–55).

◘ Table 4.3   Tallest buildings in the world in the USA 2012

Global rank	Name	Location	Height in m	Floors	Completion	Use
9	Willis Tower[a]	Chicago	442	108	1974	Offices
12	Trump International Hotel and Tower	Chicago	423	98	2009	Housing, Hotel
20	Empire State Building	New York City	381	102	1931	Offices
24	Bank of America Tower	New York City	366	55	2009	Offices
31	Aon Center	Chicago	346	83	1973	Offices
33	John Hancock Center	Chicago	344	100	1969	Housing, Offices
48	Chrysler Building	New York City	319	77	1930	Offices
49	New York Times Tower	New York City	319	52	2007	Offices
51	Bank of America Plaza	Atlanta	317	55	1993	Offices
52	U.S. Bank Tower	Los Angeles	310	73	1990	Offices
57	Franklin Center	Chicago	307	60	1989	Offices
59	JPMorgan Chase Tower	Houston	305	75	1982	Offices
63	Two Prudential Center	Chicago	303	64	1990	Offices
65	Wells Fargo Plaza	Houston	302	71	1983	Offices
75	Comcast Center	Philadelphia	297	57	2008	Offices
79	311 South Wacker Drive	Chicago	293	65	1990	Offices
83	70 Pine Street	New York City	290	67	1932	Offices
84	Key Tower	Cleveland	289	57	1991	Offices
87	One Liberty Place	Philadelphia	288	61	1987	Offices
92	Columbia Center	Seattle	284	76	1984	Offices
97	The Trump Building[b]	New York City	283	71	1930	Offices
100	Bank of America Plaza	Dallas	281	72	1985	Offices

Source: ► www.ctbuh.org
[a]Formerly Sears Tower
[b]Formerly Bank of Manhattan Building

In September 2011, Americans suddenly realized that skyscrapers can be largely protected against earthquakes, but not against attacks from the air. The U.S., which had not been attacked on its own territory since the Japanese invasion of Pearl Harbor in 1941, was in shock, and just days later the Wall Street Journal, USA Today, and the Washington Post unanimously announced the end of the skyscraper, saying they were not only dangerous but, in the digital age, a relic of days gone by. But the skyscraper is far from dead; in fact, more skyscrapers were built worldwide in the first decade of the twenty-first century than in any preceding decade. The skyscraper is predestined for the concentration of all global activities in a single location. Skyscrapers make optimal use of a plot of land, but reach the limits of efficiency when they have more than about 70 floors. As the height of the buildings increases, the cost of ensuring stability rises sharply, and additional elevators have to be built, which not only incur costs but also take up a lot of space. Therefore, the construction of extremely tall buildings cannot be justified from an economic point of view (Lamster 2011, pp. 78–80; CTBUH 2013).

4.5.11 Inner-City Housing

More interesting than the maximum height of the skyscrapers and their global distribution, however, is the change in use. For decades, skyscrapers were used almost exclusively for offices. Between 1930 and 2000, the proportion of office buildings was never less than 86%; in 2010, this figure was only 46%, i.e. more than half of

4

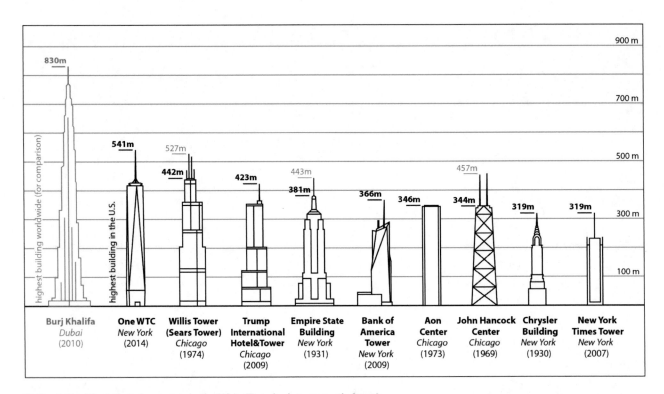

the world's 100 tallest buildings are dominated by a residential function or mixed use (CTBUH 2008, p. 41). No precise data are available for the USA, but here too this trend can be clearly traced. Well-known residential skyscrapers that have been built in recent years, some with hundreds of apartments, are the Beekman Tower (76 floors) in Lower Manhattan and the Trump Tower (92 floors) and Aqua Tower (87 floors) in downtown Chicago. Without the construction of a greater number of high-rise residential buildings in Chicago, the tremendous increase in downtown population would not have been possible. It is a concern, however, that almost exclusively apartments in the luxury segment have been built. In recent times, the construction of high-rise residential buildings has slowed down somewhat, as many were completed around 2010 and the apartments could not be sold as planned due to the financial crisis. The only way out is often to rent out the planned condominiums. In Chicago, construction of the Chicago Spire, which at 610 m was to be the tallest residential building in the world, began in 2007. Construction work was halted in the fall of 2008. The property, which is located directly on Lake Michigan, has since been put up for sale. Downtown housing has increased not only through the construction of high-rises, but also through the renovation of dilapidated structures or the conversion of former commercial spaces, which has been encouraged by various programs in many cities. Lofts are one-room apartments created by converting formerly commercial

■ **Fig. 4.22**   Loft building in SoHo, used by an expensive grocery store and a gallery

or industrial space (■ Fig. 4.22). Because partitions are absent and ceilings are high, there is ample room for individual design. The first lofts were created in New York in the 1960s, mostly by artists and other freelancers who needed a lot of space, by converting former factories and warehouses. It was not uncommon for zoning specifications to be violated, and the standard of the lofts, which often had no heating, was low. The best-known example is the conversion of former warehouses in New York's SoHo district (Zukin 1982, pp. 58–81). The pioneers of lofts created a new trend, and from the late 1980s demand increased and prices virtually exploded, as new ways of living could be easily implemented in open lofts. With an inherently limited supply of factories and warehouses suitable for conversion to apartments, investors in New York and other cities were soon erecting new buildings with large one-room apartments, which they marketed as lofts. Houston is now considered one of the prime locations for new lofts. In Chicago, investors built row houses on a parking lot and then marketed them as *row lofts*. In the copies, unused pipes are even sometimes attached to the ceilings as a reminder of a non-existent former use. A distinction is made between lofts with only one large room and *soft lofts*, in which the bedroom is separated. Unlike the "real" lofts, the copies always have large panoramic windows. Whatever one thinks of the new lofts, it is indisputable that they represent an alternative to traditional housing and are very popular among certain groups of the population (Gratz and Mintz 1998, p. 317; Postrel 2007a). Cities in the northeastern United States, in particular, have a large number of warehouses built during industrialization on the edge of what are now downtowns near waterways. As industry and commerce moved to suburban areas, these predominantly brick buildings were abandoned and often stood vacant for decades. Others have been redeveloped since the 1980s, and lofts have often sprung up here as well. In good locations, demand from new urbanites is high. One disadvantage of the lofts is the limited incidence of light, which is partly compensated for by the addition of balconies. In summer, this has created attractive places to stay, often with a good view of adjacent water areas.

Downtown Philadelphia has developed positively since the 1980s due to the increase in office and retail space, but there was a lack of housing. Potential investors argued that converting vacant office buildings and warehouses into housing was not financially worthwhile due to high construction costs and property taxes once completed. In 1999, the City of Philadelphia adopted a program that provided a 10 year tax break for conversion to housing. The more expensive the projects were, the more money investors could save. As many once prestigious office buildings and warehouses stood empty in downtown, investors focused on creating housing in those buildings. Another program provided tax incentives for building on vacant lots. After the turn of the millennium, multi-million dollar condominium sales were no longer uncommon in the downtown area. However, it was mostly large investors and wealthy buyers or renters who benefited from the tax exemptions. In addition, downtown revitalization also increased the value of older homes and property taxes. Old property owners were penalized in two senses: they did not

Fig. 4.23 Advertising for luxury apartments in downtown St. Louis

benefit from tax breaks and actually had to pay higher taxes. Philadelphia's programs to create inner-city housing through tax relief are seen as an economic success, but one that negatively impacted those who had persevered in the city during the many years of crisis and who should have been rewarded (Kromer 2010, pp. 31–47).

The downtowns were long the most important office location in the region and have a correspondingly large number of older office buildings whose layout and technology no longer meet today's requirements. Since demolition of the often high and very massive buildings is difficult, some of them have stood empty for decades. In order to revitalize the unused buildings and meet the growing demand for residential space, many cities now provide tax incentives for the conversion of offices into apartments. This is even the case on New York's Wall Street, which is arguably the world's most famous banking and stock exchange center. It is not surprising that here, too, only very expensive luxury apartments have been built, since the investors had the well-off bankers in mind as buyers or tenants. But even in the center of the rapidly shrinking city of St. Louis, new luxury apartments are being built (Fig. 4.23).

4.5.12 Architecture of Attention

As cities today are in great competition with each other, each tries to stand out from the crowd through spectacular buildings or mega-events. Expensive star architects are commissioned to create unique selling points for global marketing (Gerhard 2012, p. 57). Only when a building is featured in global glossy magazines does success set in. In 1973, the opening of the Opera House in Sydney, Australia, attracted a great deal of attention, even though Danish architect Jørn Utzon

4

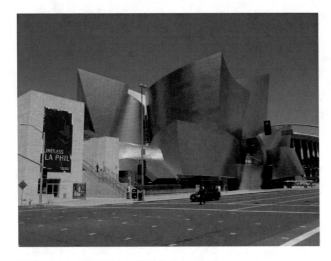

◘ **Fig. 4.24**   Walt Disney Hall in Los Angeles

was largely unknown outside his home country at the time. UNESCO has designated the opera house a World Heritage Site, and worldwide the unusual building is Sydney's most important advertising medium. The Guggenheim Museum, designed by star architect Frank Gehry and opened in 1994 in Bilbao, Spain, became similarly famous. By 2005, Bilbao's visitor numbers had risen from 1.4 to 3.8 million. This success, known as the Bilbao effect, is something that many cities are trying to emulate by hiring outstanding architects (*starchitects*) to build the most salient buildings possible (*starchitecture*) (Glaeser 2011a, p. 65; Knox and McCarthy 2012, p. 343). Los Angeles, Chicago, and Seattle also had Gehry build the Walt Disney Concert Hall on the edge of downtown Los Angeles (◘ Fig. 4.24), the oversized concert shell Jay Pritzker Pavillion in Chicago's Millennium Park, and the EMP Museum (EMP = Experience Music Project) for popular music at the base of the Seattle Space Needle. All of the aforementioned buildings reveal the signature of the famous architect at first glance. The Public Library of Seattle, built by Rem Koolhaas and Joshua Prince-Ramus, has also received much attention. Denver had the Art Museum expansion built by Daniel Libeskind, who the city of New York also hired to redesign Ground Zero. The choice of these renowned architects alone seems to guarantee global attention. However, there are only a limited number of architects whose work is recognized around the world or who are capable of producing truly spectacular buildings. The production of uniqueness is limited and expensive, and the more cities vie for global attention, the harder it becomes to achieve. Postmodern buildings also stand out because they never blend in with their surroundings, but create the greatest possible contrast. The architect Rem Koolhaas aptly summed it up with *"Fuck the context!"* (quoted in: Augé 2011, p. 128). However, a spectacular building in a neighborhood in need of rede-

velopment often does not help the residents. It attracts visitors interested in the arts, for whom expensive cafés and similar facilities are built. As a result, rents may rise and residents of the neighbourhood may be displaced (Glaeser 2011a, p. 65).

4.5.13 Quality of Stay

Vibrant city centers are more attractive to residents, employees and visitors than deserted streets and derelict shop premises. But efforts to create pleasant open-air lounges have not always been successful. In 1958, the Mies van der Rohe-designed Seagram Building was completed on the east side of Park Avenue between 52nd and 53rd Sts. The building is surrounded by a large plaza that initially received much praise. However, the idea that open spaces between buildings attract people has been proven wrong, as they are well used by smokers at best. Nearly every U.S. city in the 1950s and 1960s had similar buildings and open spaces without loosening elements such as fountains, water basins, sitting areas, flower beds, or shade trees (Owen 2009, p. 176; Whyte 1988, pp. 103–132). Meanwhile, the spaces between buildings are being designed to be much more interesting than they were just a few decades ago. It is amazing that even relatively small gaps between two high-rise buildings can become attractive amenity spaces. In the USA, these areas are often referred to as *park*, even if they only have benches, small water areas and other design elements, but not green spaces, which are an essential element of parks according to the German understanding. Nevertheless, the open spaces are well received, and employees of the adjacent offices like to spend their lunch breaks here or out-of-town tourists take a rest from their strenuous sightseeing program. The small parks have often been created as part of incentive programs and were created by the investors of the buildings standing on the same property. Since they are private areas, undesirables are at best tolerated in small numbers for short stays. However, it is advantageous that the areas are always maintained and clean (Kayden 2000).

Portland, OR, has long taken a holistic approach. Here, the paradigm shift from a car-oriented to a pedestrian-oriented city was initiated in the 1970s. The downtown area is located on the western bank of the Willamette River with a checkerboard layout typical of U.S. cities. Because the individual blocks are exceptionally small at only 61 by 61 m, the ratio of open space to built-up area is favorable. In addition, the distances from one intersection to the next are short. Thus, good conditions for an increase in attractiveness were present. In the downtown area, sidewalks were widened at the expense of roadways, signaling to pedestrians that they were the focus of attention. Guidelines have influenced the design of streets, sidewalks, and plazas, and the height of new buildings has also been adjusted to the street width to create a livable environment where pedestrians feel comfortable. Pioneer Courthouse Square occupies an entire city block in downtown Portland and is one of the most attractive and best-used squares in the United States. A civic group called the Friends of Pioneer Courthouse Square lobbied for the transformation of the square, which was previously partially occupied by a multi-story parking garage, and raised about $150 million from private and public sponsors for the early 1980s redesign. The square's 50,000 paving stones were donated by private citizens, whose names were engraved on the stones in gratitude. This provided the best possible way for Portland citizens to identify with their new square. Pioneer Courthouse Square is surrounded on four sides by department stores, hotels, shopping centers, and public buildings, almost all of which have ground floor storefronts. In the center, seating has been created on curved staircases that provide a good view of a lower stage, similar to an amphitheater. The square is used almost daily in the summer at midday for events and is always well attended in good weather (Gehl 2006, pp. 60–65, 232–235; author's observations).

In Washington, D.C., the long-neglected banks of the Anacostia River have been upgraded since 2004, and in 2011 the first section near the Navy Yard was opened to the public. An attractive riverside walkway is always complemented by small parks that offer sun loungers for rest and relaxation or water pools and fountains that are popular with children and young families (❏ Fig. 4.25). Many American cities are fortunate to have large parks that were created before all of the land was built up in the face of great settlement pressure from immigrants. Particularly noteworthy are Boston Common, which grew

out of a cattle pasture used in the seventeenth century and was expanded in the 1830s to include the Public Garden, and New York's Central Park, which was created in the second half of the nineteenth century and covers an area of almost 3.5 km². Green spaces and parks not only form green lungs in compact inner cities, but can also develop into real visitor magnets. However, it has been shown that the intensity of use of small green spaces is far greater than that of large parks. Central Park is exceptionally well used by all groups of the population and reportedly has about 25 million visitors annually. Yet the interior of the park is often deserted, as a sense of dread occasionally creeps into the more natural and sometimes almost overgrown parts of the interior. Washington Square Park, on the other hand, which is only about 4 ha in size, is much more intensively used. The park can be seen from all sides and offers a wealth of attractions in both summer and winter (Owen 2009, pp. 169–170).

Interesting parks have large water and lawn areas, but also sculptures from different eras. In some cities, inner-city parks have been created in recent years with generous donations from companies and private individuals, such as Millennium Park in Chicago or the Gateway Mall in St. Louis, which have also been enhanced by works of art (❏ Fig. 4.26). However, sculptures are

❏ **Fig. 4.25** Redesigned riparian zone of the Anacostia River in Washington, DC

❏ **Fig. 4.26** View from the St. Louis Arch to downtown and Gateway Mall

4

also often found in plazas between high-rise buildings, as they can enhance inner-city spaces. In the *Loop* of Chicago, an abstract bull skull by Picasso, about 15 m high, and a 16 m high flamingo by Calder, in the artist's typical style, had been installed in 1967 and 1974 respectively. In 1978, Chicago became one of the first American cities to adopt a *Percent for Art Ordinance*, which required that 1.3% of spending on public buildings and spaces be devoted to works of art. The guidelines have since been changed several times and many cities have similar programs (▶ www.cityofchicago.org). Chicago relies on large-scale sculptures, as small artworks between the skyscrapers would hardly attract attention. Other cities are trying to increase attractiveness with smaller measures. In downtown St. Louis, the Citygarden sculpture park was created in 2009 on a 1.2-acre brownfield site with artworks by 24 artists; starting at the State Capitol, it is an extremely attractive corridor and has greatly increased the attractiveness of downtown. The $25 million park was donated by the Gateway Foundation (Gallagher 2010, p. 106).

4.5.14 Retail Trade

With suburbanization, retail had left the inner cities and migrated to suburban areas. In the 1970s, the first shopping centers were built in downtowns, where their often windowless and forbidding facades were reminiscent of spaceships in an inhospitable environment. In the meantime, retail has developed extremely positively in many downtowns and is in strong competition with even the largest suburban shopping centers. Customers are the residents, employees and tourists of the inner cities, but also visitors from suburban areas who prefer the colourful urban life to the monotonous shopping centers. Successful downtown shopping centers have been a spark for street-level retail in some large cities, such as San Francisco and Chicago. Only when the shopping temples were clearly successful did retailers invest again on traditional shopping streets, which was initially risky and expensive because older buildings had to be refurbished or converted and customer acceptance was not assured (Hahn 2002, p. 58). However, this positive development has not taken place in all downtowns. In part, downtown retail is still concentrated in shopping centers, which can even prevent street-level retail development by contractually prohibiting their tenants from opening additional stores within a certain radius of the shopping centers (Gerend 2012, p. 119). In Phoenix and San Jose, which grew comparatively late, downtowns were never significant retail locations; revitalization is therefore not possible. In mid-sized cities, such as the former steel hub of Birmingham, AL, investments have

been made in manicured green spaces and museums, but retail has not returned. Here, either the shopping centers in the suburban area are too strong or the image of the downtown is too poor (author's observation).

Inner-city shopping centers that have been built more recently differ significantly from older centers. Salt Lake City hosted the 2002 Winter Olympics, for which the first phase of the Gateway Center opened (◘ Fig. 4.27a). This is an open center, with two levels of shops opening onto a kind of traffic-calmed thoroughfare (◘ Fig. 4.27b). Above the shops are offices and apartments. Since its completion in 2007, the complex, which also includes Salt Lake City's old train station concourse, has housed about 100 shops and restaurants, as well as a multiplex movie theater and planetarium. The Gateway Center attempts to copy the city and disguise the fact that it is a man-made shopping center planned and managed by an investor. Although this has not been entirely successful from a European perspective, the Gateway Center has led to other investments in downtown Salt Lake City such as the construction of the City Creek Center, which is also multifunctional (author's observation).

Farmers' markets are increasingly contributing to the revitalisation and supply of the inner-city population (◘ Fig. 4.28). Their number was still declining in the 1960s, because people preferred to buy in supermarkets on the outskirts of the city and the food rarely came from the surrounding area. An industrialisation of agriculture had begun. Cultivation became more efficient and cheaper, but transport distances became longer and longer. Today, even in the centers of very large cities, area farmers regularly offer their produce. They can be found in Washington, D.C., at Dupont Circle, in Boston on the edge of Boston Common, in Chicago in front of the Museum of Modern Art, and in Union Square in Manhattan. The markets are colorful, feature a rotating array of offerings, and attract shoppers and onlookers in large numbers on select days of the week. They are supplemented by indoor markets, such as those in downtown Baltimore and Los Angeles, which are frequented almost exclusively by locals, and by *festival markets* with many out-of-town visitors, such as Faneuil Hall Marketplace in Boston and Pike Street Market in Seattle. Faneuil Hall Marketplace has long been dominated by chain stores catering primarily to out-of-towners, while Pike Street Market strives to maintain its authenticity. Here, there is an interesting mix of local fish, vegetable and flower vendors, and high-end fast food vendors, as well as seafood restaurants. Only the offerings of a few stalls are aimed at tourists. Although Pike Street Market deliberately avoids advertising, it is reportedly the largest tourist attraction in Washington State (Gratz and Mintz 1998, pp. 222–223; author's observations).

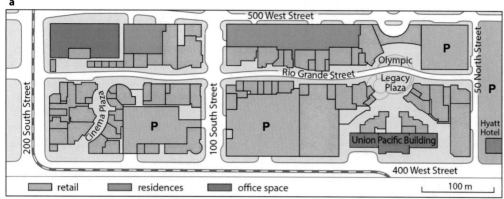

⬛ Fig. 4.27 a, b Gateway Center in Salt Lake City. (**a** Adapted from documents of the center administration)

4.5.15 Transition Zone

The transition zone connects two differently structured parts of the city, but cannot be assigned to either of them (⬛ Fig. 4.29). Land uses are different and often less orderly than in the adjacent neighborhoods. Often the development predates the introduction of land use plans. Buildings of different uses and heights therefore often share a building block. A close juxtaposition of commercial businesses, warehouses, and older residential buildings is characteristic of the transition zone (Kromer 2010, p. 77). In post-industrial cities, transition zones create a ring around inner cities and separate them from residential areas. In the nineteenth century and early twentieth century, factories and densely built-up workers' quarters had developed here. Gradually,

◘ **Fig. 4.28** *Farmers market* in downtown Boston

◘ **Fig. 4.29** Transition Zone at the Edge of Downtown Seattle

the industrial plants at this location closed down. The abandoned buildings remained unused or were used by small businesses, as repair shops or as warehouses. A low-income population, including many immigrants and members of ethnic minorities, moved into the former workers' housing. For decades, however, the transitional zone was abandoned even by these residents, and increasingly, brownfields dominated the scene. When some inner cities started to develop positively again at the end of the twentieth century, the number of inhabitants in most of the transitional zones was still declining. An upgrading of the transition zone can support the positive development of the inner cities as well as the adjacent residential areas, because there is a large supply of land for new residential and office buildings, but also parks, cultural facilities or sports facilities. In some American cities, the pressure on the inner-city real estate market has been so great that investors have moved to adjacent areas in recent years. Boston's Four Point

neighborhood was separated from downtown only by a canal and was once a manufacturing hub. Now, many of the former factories have been converted into expensive apartment buildings and Four Point has become a popular residential location. Downtown Chicago is also expanding at the expense of the transition zone. In the south end of downtown, high-rise luxury apartment buildings have been built in recent decades, and properties that have been unoccupied for decades have been redeveloped and sold at a high price. The positive development of the downtown real estate market here has "*spilled over*," so to speak, into the adjacent transition zone, where no one would have voluntarily set foot two decades ago (author's observation). Other cities have supported the revitalization of the transition zone through incentive programs. Philadelphia established a "*Home in Philadelphia*" program that provided tax support for housing. New neighborhoods with loose development sprang up to provide near-city alternatives to suburban space. Adherents of *new urbanism*, however, would have preferred more compact uses that conserved land (Kromer 2010, pp. 77, 87). This positive development of the transition zone cannot be observed in all cities, however, because often, as in Detroit, the demand for real estate is very low and, in addition, a ring of freeways around the downtowns prevents this process.

4.6 Segregation

The population can be assigned to different groups based on their ethnicity, socio-economic or demographic characteristics. It is not uncommon for certain groups to be concentrated in individual neighbourhoods of the city and for there to be an uneven distribution of population across space. This phenomenon is known as segregation. Segregation can be the result of a voluntary or an involuntary process. If people with similar lifestyles or the same ethnicity prefer a particular neighborhood, there is nothing wrong with it as long as they do so voluntarily and other groups are not excluded. Segregation has always existed in U.S. cities, and almost every ethnic group has concentrated, at least temporarily, in certain neighborhoods. Designations such as Little Sicily, Little Italy, Japantown, Little Havana, or Chinatown indicate a strong presence of an ethnic group in a particular neighborhood, even if the members of the ethnic groups have long since moved to another location. The repeated replacement of the entire population in a larger district or neighborhood is an important characteristic of the U.S. city.

Cities do not drive people into poverty, but cities attract poor people in search of work. Thousands of poor people from around the world continue to settle in major American cities each year in the hope of a better life (Glaeser 2011a, pp. 9–10). First-generation poor immigrants have lived in crowded neighborhoods in grossly inadequate sanitary conditions, which at the same time developed great momentum as all residents sought to improve their living conditions. The crowdedness and diversity of residents fostered the emergence of social tensions and even violence between new immigrants and longer-time residents and between Protestants and Catholics (Muller 2010, pp. 316–318). The North End in Boston, where the Irish settled in the mid-nineteenth century, or the Lower East Side in New York are examples of once highly segregated neighborhoods. The Lower East Side, located in Manhattan between Canal St., Houston St., the Bowery and the East River, was the gateway for the Irish in the mid-nineteenth century, and from 1880 onwards the Germans arrived, and soon Eastern European Jews in large numbers, and also Italians and Poles. As Chinatown expanded, Chinese also moved to the Lower East Side. In 1915, 320,000 people lived here, and the Lower East Side had one of the largest population densities in the United States. When it came to choosing a profession, ethnicity was crucial. The Irish provided the policemen, the Germans brewed beer, and the Chinese maintained laundries. The Lower East Side also developed into an early center of the textile and garment industry in New York. It was only a few years ago that the neighborhood was gripped by *gentrification*, making it attractive to more affluent citizens (Roman 2010, pp. 129–131). In the 1970s, Hispanics from Mexico, the Caribbean, and Central America, as well as Asians, began arriving in greater numbers in the cities of the Southwest, where, like members of earlier waves of immigration, they settled preferentially in the central neighborhoods. They took low-paying jobs while building their own networks and institutions. Meanwhile, hispanics are also found in many Northeastern cities, where they are also concentrated in certain neighborhoods. The proportion of hispanics on Chicago's Lower West Side is 82%, and East Harlem in northeast Manhattan, also called Spanish Harlem, is one of the largest concentrations of hispanics in the United States (▶ www.census.gov). Puerto Ricans had settled here after the First World War and laid the foundations for the influx of further hispanics.

4.6.1 Ethnic Segregation

At the beginning of the twentieth century, hardly any blacks lived in the cities. In 1900, this applied to only 2% of New York's residents and 1.8% of Chicago's residents. In the following decades, more and more blacks moved from the South to the industrial cities in the Northeast in search of jobs. The mostly destitute immigrants were forced to concentrate in the poorest neighborhoods, which were promptly abandoned by whites, not infrequently creating ghettos with an almost exclusively black population. In addition, cities designated neighborhoods into which blacks were not allowed to move (*ethnic zoning*). Although the National Association for the Advancement of Colored People (NAACP) successfully challenged this practice in the U.S. Supreme Court in 1917, municipalities rarely intervened when so-called undesirables were excluded from renting or buying. States supported racial segregation in the housing market by making it more difficult to obtain mortgages in ethnically mixed neighborhoods and banning blacks from moving into certain neighborhoods. These so-called *restrictive convenants* were not overturned by the Supreme Court until 1948, but still not all states complied. It was not until 1964, when the United States Congress finally abolished discrimination against blacks and racial segregation with the *Civil Rights Act*, that the disadvantageous treatment of certain groups of the population by banks or politics was successfully eliminated (Freund 2007, p. 207; Glaeser 2011a, pp. 82–83; Gotham 2000, pp. 618–619) (◘ Fig. 4.30). By this time, however, the white population was moving in large numbers to suburban areas. Cities lost their economic base and blacks lost their jobs. The result was the spatial separation of the poor blacks, who remained in the core cities marked by decay, from the affluent and predominantly white middle class, who had good access to the real estate market and jobs in the suburban area (Freund 2007, p. 6).

Few blacks could afford to move into suburban areas, where they then concentrated in just a few neighborhoods. One of these exceptions is Palmer Woods, which lies north of Seven Mile Road and west of Woodward Avenue in Detroit, where it forms an enclave. In 1915, the Palmer family had purchased 76 acres of

◘ **Fig. 4.30** Street band of black musicians in Washington, DC

4

■ **Fig. 4.31**   Single-family house in Palmer Woods

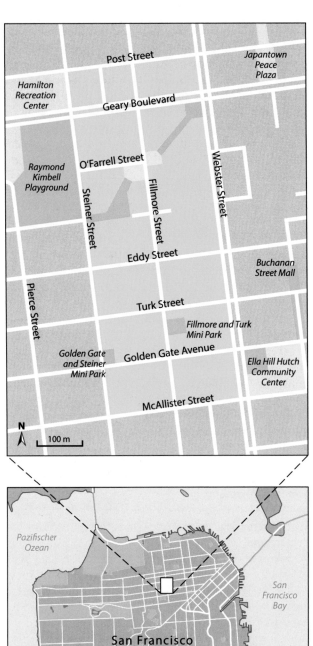

■ **Fig. 4.32**   Fillmore District in San Francisco (Lai 2012, Fig. 2; heavily modified)

land here. Landscape architect Ossian Simonds laid out a neighborhood of curving streets and irregularly shaped lots to disrupt Detroit's checkerboard pattern and avoid any impression of uniformity. Street names such as Gloucester, Balmoral, and Cumberland recall British history. Many industrialists, seeking to escape the noise and grime of the nearby big city, built lavish mansions here in the style of European aristocrats, complete with libraries and separate entrances and staircases for servants (■ Fig. 4.31). The houses were built in the style of the British Elizabethan and Jacobean periods (1588–1625), known in the United States as the *Tudor revival*. Among the architects of the sometimes palatial estates were such well-known names as Frank Lloyd Wright and Albert Kahn (Palmer Woods Association 2011). The builders stipulated that the exclusive homes were to be sold exclusively to whites. After *restrictive covenants* were banned in 1948, a few successful blacks initially moved to Palmer Woods, but as whites increasingly left the community, more and more blacks came. For decades, Palmer Woods, which includes nearly 300 homes, was considered the most successful black-inhabited community in the United States. Some 81% of residents are now black, and household incomes are above average. It is only in very recent times that the negative influence of Detroit seems to be having an impact on the Palmer Woods enclave. Crime rates are on the rise and home prices have dropped significantly (Coates 2011).

The Fillmore District in downtown San Francisco, which occupies about 50 blocks west of Van Ness Avenue between California St. and Geary Blvd. provides a good example of the intensification of contrasts through urban planning and the transformation that many neighborhoods have experienced within a few decades (■ Fig. 4.32). Until the 1906 earthquake, Jews had lived preferentially in the Fillmore District. Because

the neighborhood suffered comparatively little destruction, it took over many of the functions of the adjacent downtown, and population densities rose sharply. Beginning in 1910, thousands of Japanese moved in, and the Fillmore District became known as Japantown. After the Japanese attack on Pearl Harbor in late 1941, the Japanese fell out of favor and were interned in camps for several years (Raeithel 1995, vol. 3, pp. 146–147). Blacks and whites, many of whom worked at the Port of San Francisco, moved into the empty apartments, which were often of poor construction. The Fillmore District developed into the most important center of African American music. Famous artists such as Ella Fitzgerald, Louis Armstrong and Billie Holiday performed in the numerous nightclubs. Since the Fillmore District was considered one of the most run-down neighborhoods in downtown San Francisco after World War II, it was declared a redevelopment area in 1948, which at the time primarily meant area redevelopment, i.e. demolition and new construction. In the 1970s alone, some 5000 affordable housing units were destroyed and 13,500 residents lost their homes. Mostly low-rise apartment buildings and multi-purpose buildings were rebuilt, and public funds were used to construct social housing along Fillmore St. The low-cost public housing was sought after by working-class blacks and whites, as well as Japanese who were pushing back into their ancestral neighborhoods. At first, both blacks and Japanese were considered unpopular; but over the decades, as Japan's political and economic importance grew, the status of the Japanese changed. The city increasingly supported the construction of facilities for Japanese in the Fillmore District. The old Japantown designation reappeared on official maps and plans, which did not sit well with the other residents of the neighborhood. But even the Japanese were not unqualifiedly enthusiastic about the rebuilding, for some of the new buildings were built as white Americans imagined Japan to be. A *disneyfication* took place, and Japantown developed into a popular tourist destination. However, the Japanese themselves contributed to the marketing of the neighborhood through the sale of Japanese arts and crafts, Japanese restaurants, a Japanese Cultural and Trade Center, and a Miyako Hotel (🔲 Fig. 4.33). People pursued commercialization and exploitation in imitation of Japanese motifs, although the Fillmore District was an increasingly rare residential location for Japanese people. In the past 20 years, there has also been *gentrification* of the neighborhood with extensive population replacement. The process of *gentrification* is well underway, and white and black longshoremen have been replaced by well-off whites and Asians (Lai 2012).

Better known than the Japantowns are the Chinatowns, many of which were established as early as the mid-nineteenth century, when the first Chinese were attracted by the gold discoveries in California and shortly afterwards became important workers in railway construction. The immigrants, almost all of them male, planned to return to China to their families as soon as possible. In the cities, they formed segregated enclaves

🔲 **Fig. 4.33** Marketing of Japantown in San Francisco

with their own economies in the form of Chinatowns, which sealed themselves off against insight and outside influences. Many of the Chinatowns are still the first place of residence for Chinese immigrants today, as they can retain their culture and language here. Other Chinatowns, such as the one on the edge of downtown Washington, D.C., have developed into *entertainment districts* in which only the characteristic archways still remind us of the former inhabitants.

4.6.2 Socio-Economic Segregation

In cities, the poor and the rich have always lived side by side in close quarters in isolated worlds. Chicago's Near North Side has long been a neighborhood of extremes, juxtaposing not only very different socioeconomic groups but also building blocks with the highest and lowest land prices in the city. The Near North Side stretches 2.4 km long and 1.6 km wide west of Lake Michigan and is bordered on the south and west by the Chicago River. In the 1920s, some 90,000 people from 29 nations shared this narrow space. Most of the politicians and almost all of the 400 families who belonged to the city's Social Register (a kind of directory of prominent and wealthy residents) lived east of State St. directly on the lake on the so-called Gold Coast. Their lives followed precise social rules. In the blocks west of State St. extreme poverty spread. Typical were the *rooming houses* (cheap hotels), because many inhabitants were passing through and lived here only temporarily. Because the crime rate was very high, the neighborhood was known as "*Little Hell.*" The wealthy never crossed State St. to frequent the poor streets (Zorbaugh 1929).

Today, most U.S. cities are broadly divided into neighborhoods with large concentrations of poverty and wealth. In Chicago, the wealthy prefer to live on the North Side and the poor are concentrated on the South Side; in Manhattan, the better-off live in the South Side and the low-income live in the North; in Washington, the wealthy are concentrated in the Northwest sector of the city and the poor live south of the Anacostia River. The socioeconomic distribution of the population often correlates with ethnic segregation. The small-scale contrasts may not be as stark today as Zorbaugh depicted for Chicago in 1929, but they are still evident. Until the mid-twentieth century, members of minority groups were prohibited by law from moving into certain neighborhoods. Although these laws have long since been repealed, no real mixing of the various groups occurred, as access to a residential location is regulated by land value (◘ Fig. 4.34). Asians have by far the highest household income at around $65,000 per year, followed by whites at $52,000. These two groups are relatively free to choose where they live, while Hispanics and blacks have little choice, with household incomes of just under $39,000 and just over $32,000, respectively (► www.census.gov). Since the residents of disadvantaged neighbourhoods cannot afford to move to a better neighbourhood, they remain in their long-standing place of residence, even if it offers no opportunities for advancement or is dominated by gang crime.

◘ **Fig. 4.34**   Real estate for the super-rich

A particular problem are the large areas of concentrated poverty with high unemployment, poverty, gang crime, drug abuse, high birth rates and a large proportion of single mothers. In terms of construction, these *hyperghettos* are characterized by a great deal of decay and a lack of public infrastructure. Even those who have jobs often cannot support their families and are considered *working poor* (Schneider-Sliwa 2005, pp. 148–149). All anti-poverty programs have failed so far. Wilson (1987, pp. 109–124) referred to the poor of the core cities as the "*truly disadvantaged*" or "*ghetto underclass,*" which disproportionately includes blacks. Unlike whites who lost their jobs in the wake of deindustrialization, they were far less likely to find new employment and were not as likely to move to suburban areas (Rothstein 2012). To date, their situation has barely improved, and the demolition of many *public housing projects* has actually worsened access to affordable housing. Wacquant (2008, p. 1) deals with the black population of the USA living in the *hyperghettos* of big cities and even refers to the losers of American society as "*urban outcasts.*" According to him, politicians, the press, and other voices stigmatize the poor and their neighborhoods, calling them *lawless zones, problem estates, no-go areas,* or *wild districts. Indeed,* residents of the "*better*" neighborhoods extremely rarely or never venture into the *hyperghettos* and are close to fainting fits when foreign visitors drive or even walk into the poor neighborhoods (author's observation). These reactions show that urban populations in the United States live in different worlds. Mike Davis (1992a, pp. 265–322) has examined the influence of different social groups and ethnicities on the future development of Los Angeles and predicted radicalization in the face of increasing competition for resources. The antagonism between members of the underclass, consisting of new immigrants and the *working poor,* and the socially and economically established residents of the city would grow, and thus also the potential for conflict, which would sooner or later be discharged violently. Although since the 1990s the contrasts have intensified rather than diminished in view of lower social benefits, the predicted radicalization has surprisingly not yet occurred. It is possible that the poor of the *hyperghettos* are so preoccupied with daily survival that they have no strength left for larger uprisings.

4.6.3 Demographic Segregation

Due to higher birth and immigration rates, the US population is ageing more slowly than the German population, but here too the proportion of older people in the total population is increasing. In 2000, just over one in seven Americans was 65 or older; by 2030, this will apply to one in five Americans. Demographic segregation has increased in U.S. cities in recent decades, with rarely more than two generations living under one roof. The *empty nesters* (couples whose children have left home) are moving into *retirement communities* or staying in the former location. The neighborhoods age with their residents and become de facto *retirement communities* (NORC = *naturally occurring retirement communities*), although they were not planned as such. In 2004, a good third of households with seniors had lived in the same house for at least 31 years. Many of these neighborhoods were built in the 1950s–1970s and were not built to accommodate the elderly. At the same time, newly developed subdivisions are often occupied exclusively by young families (Kirk 2009, p. 115).

4.6.4 Segregation in More Recent Times

Although the U.S. census has regularly collected small-scale data on ethnicity since 1890 and trends can be traced over 120 years, it is controversial whether ethnic segregation in American cities has increased or decreased in recent decades. Most studies focus on the residential locations of white and black Americans, who embody the two poles of U.S. society because of contrasting skin color and incomes. The dissimilarity index measures the extent of spatial differences between black and white populations. The higher the index, the greater the segregation. This is also true for the index of isolation. Segregation and isolation increased with the influx of blacks into cities in the first decades of the twentieth century, but have declined significantly since 1970 (◻ Fig. 4.35) (Glaeser and Vigdor 2012, pp. 6–7).

Whereas 50 years ago one-fifth of all urban neighbourhoods did not have a single black resident, in 2010

this was true for only one in 200 neighbourhoods. In 1960 almost half of all blacks lived in ghettos in which they accounted for more than 80% of the population; in 2010 this was true for only 20% of all blacks. A long-term study examined 658 regional real estate markets in the 85 largest metropolitan areas and found that, except for a single real estate market, the segregation index in 2010 was below that of 1970. Segregation also declined in 522 of these markets between 2000 and 2010. This is true for large and small cities alike. Among the ten largest cities, ethnic segregation is highest in Chicago, New York, and Philadelphia, but still declining. It is particularly low in the Texas cities of Dallas and Fort Worth. Today there are hardly any neighborhoods where not a single black person lives. This trend can be attributed to the repeal of laws that encouraged racial segregation, greater access to mortgages, and free housing choice for all ethnic groups. More recently, many blacks have left traditional inner-city residential locations, where they often made up more than 90% of the population, to move to less segregated suburbs or sunbelt cities. In addition, many cities demolished social housing such as the Pruitt-Igoe complex in St. Louis or the Robert Taylor Homes in Chicago, which were almost exclusively occupied by blacks. In light of these positive results, Glaeser and Vigdor (2012) have heralded the imminent end of segregation, though this has been sharply criticized by Richard Rothstein (2012), who directs the Institute on Law and Social Policy at California University at Berkeley, who argues that the decline in black segregation is largely due to the influx of poor Hispanics and, in some regions, poor Asians into what were once all-black neighborhoods. It is also important to remember that poor and affluent blacks are far more segregated today than they were in the 1970s (Reardon and Bischoff 2010, p. 25). The end of segregation still seems a long way off.

Poor schools and low levels of education among the population continue to be a major problem in the *hyperghettos*. Many of the public schools in the core cities are considered unsafe even by teachers, and students are described as *"monsters created by poverty and racism"* (Winters 2010, p. 2). *Charter schools*, which are funded by taxpayers but largely exempt from school board mandates, represent a promising approach to addressing this problem. In New York, it has been possible to experiment independently with new forms of instruction since 2004. In Harlem, Promise Academy makes great demands on teacher and student performance. After the first year, nearly half of the teachers were fired for not meeting the goals. There are now about 100 *charter schools* in New York's five boroughs, whose success is not based on spectacular new teaching methods or small classes, but which make great demands on student learning and discipline and make the best use of available resources. Even the smallest infractions of school rules are not

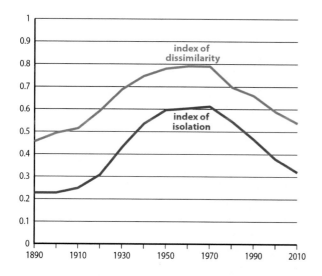

◻ **Fig. 4.35** Segregation of black and white population 1890–2010. (Adapted from Glaeser and Vigdor 2012, Fig. 1)

4

tolerated and are severely punished. Because students know the opportunities *charter schools* offer them, rule violations are rare. Students and teachers feel comfortable at the schools and can concentrate on the lessons. Unlike other public schools, there is little cost for damage caused by students. Students perform very well on tests, achieving scores comparable to students in affluent suburban communities. One criticism is that access to schools is limited and decided by lot. Chance decides the further course of life. Proponents of the system see themselves on the right track and hope that the lottery system will 1 day become superfluous and that places at *charter schools* will be available for all children (Glaeser 2011a, p. 88; Winters 2010). *Charter schools* also exist in other American cities, but even in New York only about 3% of all students attend the privileged schools, which are thus only a drop in the bucket. However, it is to be feared that the well-educated will leave the ghettos after finishing school and will not contribute to their positive development.

4.7 Gentrification

In simple terms, *gentrification* refers to the upgrading of residential areas through the influx of the middle class with the simultaneous displacement of the local population and should not be used as a synonym of reurbanisation (■ Fig. 4.36). Although there are areas of overlap and the processes are largely parallel, *gentrification "should be seen as a specific manifestation and concomitant of reurbanization"* (Brake and Herfert 2012, p. 16).

The term *gentrification* was coined in 1964 by the British sociologist Ruth Glass in reference to *gentry* (landed *gentry*) to describe processes of change in central London, where working-class neighbourhoods had

gradually been occupied by the middle classes, once shabby flats had been redeveloped and the original residents displaced. What initially appeared to be a one-off development in Islington has become a worldwide phenomenon and an important urban planning tool. *Gentrification* has taken very different forms, and the players have changed too. In 1960s London, *gentrifiers* were members of the middle class who were not afraid of living in close proximity to the lower class. Today, public-private partnerships and national and international companies in search of locations with optimal investment opportunities and high returns are important actors in *gentrification*, which, under the influence of international capital flows and liberal policies, has become a global strategy for upgrading dilapidated neighborhoods. However, *gentrification* is one of the most controversial processes in modern urban development (Smith 2006, pp. 191–193).

Gentrification had existed long before Ruth Glass made her observations in London and coined the term. Between 1871 and 1914, the nouveau riche of the city of Chicago had moved their residences from Prairie Avenue, south of the *Loop*, to the north of the city on the shores of Lake Michigan, displacing the resident gardeners (Zorbaugh 1929). Today, the so-called Gold Coast is still a preferred residential location with very high real estate prices and is referred to as *paleo-gentrified* by Conzen and Dahmann (2006, p. 9) due to its very early *gentrification* in reference to the Earth Age (■ Fig. 4.37). In addition, by the 1930s, certain neighborhoods in New York, New Orleans, Charleston, and Washington, D.C. (Georgetown), there had been *gentrification* of the built fabric with simultaneous displacement of the original residents. In the 1960s, *gentrification* began on a larger scale in Chicago and other U.S. cities. Dupont Circle and Capitol Hill in Washington, D.C., the South End in Boston, Lincoln Park in Chicago, Inman Park in Atlanta, Haight-Ashbury and The Castro in San Francisco are among the best-known *gentrified* neighborhoods in the USA. Accordingly, scholarly interest in the restructuring processes has increased. In the 1970s and early 1980s, research focused on empirical studies. Data were collected on the age, income, occupation, lifestyle, and ethnicity of the *gentrifiers*. With few exceptions, the focus was on new residents rather than old residents. Since the reasons seemed obvious, the focus was on the effects. *Gentrification* was soon seen as a panacea for decaying inner-city neighborhoods, especially as cities benefited from higher property tax revenues. The term *brownstoning* became synonymous with *gentrification* because in New York older buildings made of sandstone (*brownstones*), which become darker and redder over time, were particularly popular with *gentrifiers* (Lees et al. 2008, pp. 5–6). From the late 1970s onwards, first in the UK and somewhat later in the U.S.,

■ **Fig. 4.36** Incipient *gentrification* in downtown Philadelphia

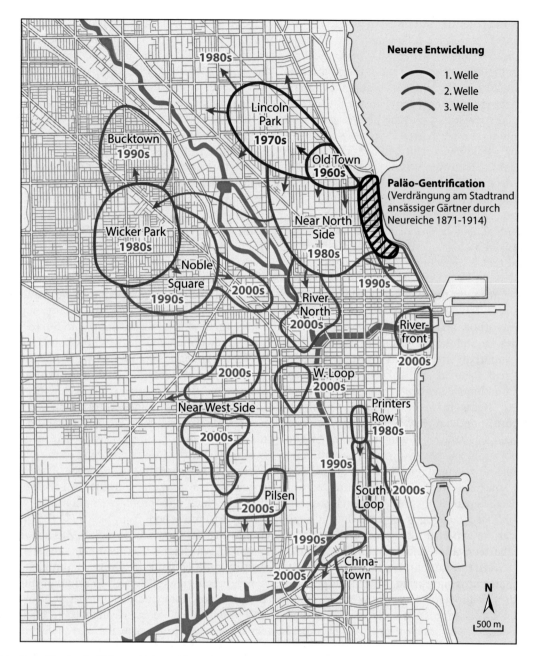

Fig. 4.37 Gentrification in Chicago. (Adapted from Conzen and Dahmann 2006, p. 9)

people also asked why the restructuring processes were taking place and developed the first theoretical explanations (Smith and Williams 1986, p. 2). At the same time, the process of *gentrification* became increasingly complex and it was found that causes and effects differed significantly in individual locations. While early definitions leaned heavily on Ruth Glass, as the process diversified, explanatory approaches and definitions became more diffuse. In the course of the 1980s, it became clear that *gentrification* was not limited to the upgrading of the building fabric, which had long been characterized by processes of decay, and the replacement of the

population in individual *neighborhoods*, but led to a far-reaching economic, social, and spatial restructuring (Lees 2003, p. 2490).

Clay developed a five-phase model in 1979 that emphasized the community influence of *gentrification* and considered population replacement and the impact on the renewal of deteriorated neighborhoods. In the first two phases, change is initiated by public administration. This may include the demolition of social housing estates or the expansion of hospitals or universities with government funding. In both cases, the resident population is relocated. In the third phase, private inves-

4

tors recognize the potential and invest in houses in need of redevelopment, and in the following phase, land use plans are changed to allow a wider range of commercial uses. At the same time, more and more long-time residents are leaving the neighborhoods because they can no longer afford the rising ground rents. In the fifth phase, the redeveloped streetscapes may be placed under historic preservation or other barriers to change, and further private or public investment will continue to drive up property values. However, it is also conceivable that a process of decay will now begin again. Both lead to displacement of the population (Clay 1979, pp. 31–32). Neil Smith (1982, p. 151) has likened decay and development of individual neighborhoods to the movements of a seesaw that is in a constant state of up and down. A neighborhood that is in decline creates the opportunity for a subsequent upswing. Smith (1986, pp. 22–24, 1987) also made a connection between capitalism and urban development and developed an explanatory approach that is still important and uncontroversial today. For several decades, capital had flowed away from the core cities to the suburban areas, where most of the new construction had taken place, while in the cities the old fabric of buildings deteriorated and land rents fell (deinvestment). Rents in the dilapidated neighbourhoods were correspondingly low. Smith (1986) refers to the gap between this and potentially higher rents after property renovation as the *rent gap*. The difference between actual and potential rents provides an incentive for investors, since in a free market economy everyone strives to make the greatest possible profit on a property. Buying a dilapidated building is worthwhile if a handsome profit can be expected after paying all the costs incurred for the renovation and the mortgage. This is especially true when the economy is performing well and demand for housing is increasing. Given the plethora of explanations, Beauregard described *gentrification* as a messy and complex process as early as 1986. Today, *gentrification* describes a variety of upgrading processes not only in urban but also in suburban and even rural areas (Lees 2003, p. 2490) and has become an economic, cultural, political, social and institutional phenomenon (Lees et al. 2008, p. 3). There are now countless case studies and a large number of theoretical explanations, and it seems that not only the process of *gentrification* but also the term itself is highly contested.

4.7.1 Gentrifier

The process of *gentrification* began before cities started to attract higher-income residents and is closely linked to deindustrialisation and the associated increase in high-end service workers, who often earn very good wages (Osman 2011). Typical *gentrifiers* were early identified as young one- or two-person households with high incomes earned in the financial sector. They sought the exciting life of the city and wanted to live close to work. Since city centers had schools with poor reputations at best, families were not among the new residents (Beauregard 1986, p. 37). Empirical studies show, however, that *gentrification* can be triggered by very different groups. In the 1970s, the City of New York allowed artists to move into vacant warehouses in SoHo and repurpose the warehouse space into lofts. The city supported the process of conversion, but did not fund it. At the time, the ground-floor commercial spaces were still home to many small businesses. The individual floors of the four- to six-storey nineteenth-century warehouses were consistently not divided into several rooms, but offered a single space of around 250 m² with high ceilings and large windows. The lofts were attractive to artists and inexpensive because of the large supply. In the 1980s, the first art galleries moved to SoHo, and by the 1990s they dominated the neighborhood. Since the late 1990s, the galleries have been displaced by the boutiques of American and international designers such as Ralph Lauren and Burberry, who could pay higher rents. Even though SoHo was no longer affordable for newcomers, it strengthened the city's creative potential and global image. Emerging young artists settled in other neighborhoods like the Lower East Side or the Meat Packing District, ushering in *gentrification* here. Larger exhibition spaces were also available in the former slaughterhouse district, which also prompted many of the established galleries to move here from SoHo (◘ Fig. 4.38) (Gratz and Mintz 1998, pp. 295–298, 308; Zukin 2010, p. 238).

Brooklyn Heights was laid out in the early nineteenth century on a cliff on the banks of the East River with a very good view of Lower Manhattan and was long the preferred residential location of wealthy merchants and businessmen. As increasingly less affluent commuters moved to Brooklyn in the early twentieth century, the upper class left Brooklyn Heights and the fabric of the building deteriorated visibly. *Gentrification* was initiated in the 1940s when writers such as Arthur Miller, Thomas Wolf, and W. H. Auden moved here (Lees 2003, pp. 2491–2491). After the construction of the controversial Cadman Plaza West was approved in Brooklyn, the city had taken steps to protect the adjacent blocks, establishing New York's first *historic district* here in 1964. In the following years and decades, more and more Manhattan residents moved to Brooklyn, which offered a good alternative between the noisy and expensive Manhattan and the suburban space far from traffic (Osman 2011). In the 1990s, a new wave began, which Lees (2003) calls *super-gentrification*. The new generation of *gentrifiers* are in many cases bankers who pay

Fig. 4.39 The Castro in San Francisco

Fig. 4.38 Gallery in the Meat Packing District

cash for the houses with the annual bonuses and subject them to a noble renovation to a previously unknown extent. Money no longer seems to matter. They are often relatively young buyers who grew up in other parts of the U.S. or in suburban areas and are unaccustomed to New York's cramped living conditions. They convert houses that were previously occupied by multiple parties into single-family homes and replant the gardens along suburban lines. Old trees and shrubs are replaced by monotonous lawns with oversized barbecues. Within a few years, real estate prices and incomes of Brooklyn Heights residents have reached astronomical heights. In *super-gentrification*, neighborhoods with no signs of structural decay and no socioeconomic deficits experience further *gentrification*, displacing the pioneers of *gentrification*. There is no end in sight to the process (Lees 2003). In other locations, homosexuals initiated *gentrification* as they moved to urban centers early on to escape the suburbs dominated by families (Lees et al. 2008, p. 234). In San Francisco, homosexuals *gentrified* the working-class neighborhood of The Castro in the 1960s and 1970s and developed it into a symbol (◱ Fig. 4.39). Other neighborhoods where homosexuality, bisexuality, and transsexuality are very openly displayed include New York's West Village and

Miami South Beach. In Haight-Ashbury, also located in downtown San Francisco, hippies were involved in *gentrification*. A counterculture had developed out of the protest movement during the Vietnam War (1964–1973). While many people sought a better life in rural or suburban areas, others moved to the cities to develop a new lifestyle. San Francisco had the most appeal for hippies. Today, there are still shops and cafes in Haight-Ashbury that reveal the lifestyle of the flower children. But some of the hippies rose to become entrepreneurs. Perhaps they sold drugs first, then psychedelic posters, then used clothing, and eventually higher-end products. They redeveloped and sold their homes or stayed there because they appreciated the still colorful neighborhood. The local population left the district because they did not like the lifestyle of the hippies or because the location became too expensive (Zukin 2010, p. 15).

Because *gentrifiers* have different consumption habits and use cities differently than previous residents, the locales and businesses are changing as well. In some neighborhoods, artists, software developers, writers, and musicians sit at brunch in the late afternoon and go to work at midnight or relax in clubs that have been set up in derelict industrial buildings. Others have long since dropped out of the earning process and shopping has become the main occupation. The original residents suffer from the loss of identity and authenticity of their neighbourhoods, by which originality is meant. Zukin (2010, pp. 4–9) suggests preserving this by maintaining the historic fabric and keeping small shops and cafes. The *gentrifiers* encourage the alienation process by readily giving the *gentrified* neighborhoods new "*fancier*" names such as Boerum Hill, Gobble Hill, or Caroll Gardens in Brooklyn. Problematically, they influence city policy to their liking. On the site of the former Atlantic Yards in Brooklyn, they were able to push through the construction of a basketball stadium, while the lower-income population preferred the location of new industries with a large supply of jobs on the brownfield site (Osman 2011).

4.7.2 Public Funding

Gentrification was initiated in its early stages not only by individuals but also by the public sector. In 1974, the *Federal Urban Homesteading Program* was passed in order to stabilize endangered neighborhoods and create housing for the less affluent. Under the program, vacant homes were sold for a very low price, in some cases as little as one U.S. dollar, on the condition that the buyers would renovate the home and live there themselves for at least 3 years (Lees et al. 2008, p. 9). Baltimore has also developed a *Healthy Neighborhoods Program* that facilitates the purchase of homes in deteriorated neighborhoods that still have some potential through reduced mortgages. In addition, minor upgrades have been implemented to boost neighborhood confidence and increase home values (The American Assembly 2011, p. 10). In the downtown neighborhood of Fells Point, which lies northwest of the Patapsco River and had long been the traditional home of longshoremen, the program was extremely successful. The houses soon became very popular with prospective buyers and property prices rose. At the same time, the neighborhood experienced social stabilization and the living environment improved. The price, however, was the displacement of the original population. State and local programs had inadvertently initiated the *gentrification* of Fells Point. More recently, the potential of *gentrification* for urban development has been recognized and targeted for support. Cities are working with the private sector to facilitate the process. Dilapidated *neighborhoods* represent a reservoir of poverty and, unfortunately, often crime. Because municipalities have a vested interest in *gentrifying* unattractive neighborhoods, *gentrification* evolved into a planning tool used to economically, physically, and socially upgrade dilapidated neighborhoods. This process is referred to as *state-led gentrification* or *third-wave gentrification* (Gerhard 2012, p. 73; Lees 2008, p. 2454; Lees et al. 2008, p. 178). Moreover, *gentrification* is promoted by global capital in search of high returns. International investors have purchased thousands of apartment buildings with low-cost housing, redeveloped them, and sold them for high rents as condominiums (Zukin 2010, p. xi). Construction companies have thus taken over the former role of individual *gentrifiers*. In some cases, there is even talk of *gentrification* through the construction of new buildings (Lees 2003, p. 2490). However, the construction of new houses does not con-

Fig. 4.40   Mission District in San Francisco

tribute to an upgrading of the existing building stock, which is an essential element of *gentrification*. However, building on brownfield sites can increase interest in neighbouring properties and lead to higher rents, indirectly displacing the resident population (Davidson and Lees 2010, p. 408).

San Francisco is characterized by strong socioeconomic contrasts. While the wealthy always lived north of the central inner-city traffic axis Market St. (NoMa), the Mission District near the harbor south of Market St. (SoMa) was the residential area of the workers and poorer population and the location of industry and commerce (◩ Fig. 4.40). With deindustrialization, the neighborhood visibly deteriorated and became a catch-all for the city's poor. The housing market responded to the rising demand of the new urbanites by building luxury apartments and luxury refurbishment of older buildings, with a penchant for superlatives. For a long time now, the adjective luxury has not been enough for real estate agents when marketing housing; they now speak of *super luxury* or *beyond luxury*. Many of the simple and inexpensive houses were demolished and replaced by new apartment buildings. From 2001 to 2005 alone, 4000 new housing units were built, while at the same time displacing the once resident population. Recognizing the problem, the city passed the *Inclusionary Housing Ordinance* in 2002, which requires that 12% of all housing units in any new real estate project be permanently priced below the prevailing rent level in order to maintain the neighborhood's diverse population. In addition, the zoning ordinance is far more specific than in the past about preserving public concerns and accommodating all groups of the population (Cohen and Marti 2009). However, these measures are arguably only a drop in the bucket.

4.7.3 Evaluation

Gentrification can be assessed from different perspectives. There is no doubt that dilapidated and unattractive neighborhoods have been *gentrified* and that cities' property tax revenues have increased. But population replacement is a major problem for the displaced. Policymakers and planners have preferred socioeconomically mixed neighborhoods to segregated ones for decades. Whether *gentrification* increases segregation and social polarization or contributes to a more mixed population is a matter of debate. Since deteriorated neighborhoods were often characterized by severe poverty before the arrival of the first *gentrifiers*, the influx of non-local residents contributes to a more diversified population structure. However, it has been found that there is little interaction between the *gentrifiers* and the resident (poorer) population. The newcomers tend to stay away from the old residents and only socialise with their own kind, i.e. other *gentrifiers*. The *gentrifiers* have no interest in living together with members of the working class or members of other ethnic groups and prefer self-segregation to meeting unfamiliar lifestyles. Their goal is even the complete displacement of the ancestral population, which often actually succeeds in the medium to long term due to constantly rising land and property prices. As there is no evidence that socio-economically and ethnically mixed *neighbourhoods* promote community cohesion, it is questionable whether the guiding principle of the mixed city should actually be pursued further (Lees 2008). Schwarz (2010) believes that there has been too much criticism of the *gentrification* and restructuring of former working-class neighbourhoods and commercial locations. He accuses Zukin (see above) of making her observation on New York's *gentrification* only in a few small neighborhoods. New York was the largest industrial city in the world until the mid-twentieth century, and the neighborhoods studied in Manhattan were only a few blocks from New York Harbor, which employed 200,000 workers. Zukin glorifies a time when deindustrialization had already begun a profound transformation. Schwarz also criticizes Zukin's call for government funding to revitalize neighborhoods by attracting artisans and owner-operated shops, since New York has evolved from the world's largest industrial city into the capital of global capitalism.

4.8 Privatisation

U.S. cities are increasingly monitored around the clock by private security guards or video cameras. This trend was initiated by the construction of the first enclosed and air-conditioned shopping centers in the 1950s, which were at best semi-private spaces with limited access. In the 1980s, when the aim was to increase the attractiveness of city centers, which included the removal of all undesirables as well as dirt, the control mechanisms of shopping centers were transferred to many public squares and green spaces. In addition, BIDs were generously designated to increase turnover and return on land, which also involved privatizing public space. According to Mike Davis (1992a, pp. 223–226), as early as 20 years ago, the surveillance of public spaces by video cameras not only meant that no action or movement went unobserved, but also increased people's fear of these spaces. The terrorist attacks in New York and Washington, D.C., in the fall of 2011 justified the expansion of surveillance of public spaces, which, from a German perspective, sometimes borders on paranoia. A *fear economy* (Davis 2002, p. 12) has emerged that profits from citizen fear. Security has developed into an important industry, the exact scope of which can hardly be determined. According to the U.S. Department of Justice, in 2009 there were nearly 10,000 private security firms employing about one million people. That was about 80% more than in 1980, but the numbers may be much higher, as certain services are readily outsourced to subcontractors. Security guards perform a wide range of tasks such as personal protection or guarding transport, public places, shopping centers or office buildings. Doormen in restaurants or shops and gatekeepers in *gated communities* are also provided by private security firms. Estimates of the security industry's annual revenue vary from $19 billion to $34.5 billion. There are also 883,000 public sector security guards, who ideally should cooperate with private services (U.S. Department of Justice 2009). If there is only slight deviation from the norm, expulsion from *"public"* spaces by security forces occurs quickly. On a field trip with 20 students to Chicago in 2011, led by the author, hardly a day went by when we were not asked to leave our supposedly public location immediately, sometimes several times, in a more or less friendly manner, because groups (of suspicious-looking youth?) were unwelcome. What is at best annoying for an excursion is a real problem for the numerous homeless people, because for them only the dingy corners of the city remain. At the same time, more and more people are entrenching themselves in *gated communities*, sending their children to private schools, and spending their leisure time at private golf and tennis clubs or even ski resorts that are not open to the public (Frank 2007, pp. 34–80). They rarely care about the problems of the less well-off. The U.S. has become a two-class society, and this is reflected in spatial terms as well. The wealthy decide in what way and by

◘ Fig. 4.41   Filming at Pershing Square in Los Angeles

4.8.1 Shopping Centers

In 1956, the opening of the first covered and air-conditioned shopping center, Southdale, near Minneapolis, marked a turning point. The shopping center, designed by the German-born architect Viktor Gruen, was intended not only as a business center, but also as an experience for the people of Minneapolis. Flowers bloomed year-round in the atrium, which was covered by a glass roof, and concerts and other events were held here regularly (Gruen 1973, p. 22 f.). Gruen probably had no idea of the development he had initiated with this concept. Southdale was a great success, and soon similar shopping centers were being built all over the country, performing more and more functions such as a large gastronomic offer, services of all kinds from the post office to the dentist to the tax consultant, cinemas and other leisure facilities. Life without shopping centers has long been unimaginable for the majority of Americans. Teenagers spend much of their leisure time here, and older people gather for morning mall walks before the stores even open (Kowinski 1985). In suburban areas today there is often no alternative to shopping centers, even if not all of them are covered and air-conditioned. In the inner cities they have also become established and not infrequently caused a bleeding out of the traditional shopping streets. Shopping centers try to copy the colourful hustle and bustle of urban life with all available means such as light incidence, large mirrors, groups of trees, fountains and the constant movement of vertical means of transport. But it is only an imitation of real urban life. The focus is always on increasing sales. Nothing happens spontaneously, everything is planned. At the same time, all negative phenomena of real life are blocked out. This includes not only the current weather and road traffic, but also certain groups of the population. Homeless people, poorer-looking old people and larger groups of young people are usually politely but firmly shown the door. Nothing and no one is supposed to stop customers with purchasing power from spending money undisturbed. In every shopping center, a notice indicates that it is under private management and that visitors must abide by the house rules. These prohibit certain behaviour which, although not always welcome on the streets, is tolerated. In shopping centers, it is usually forbidden to sit on the floor or behave in any conspicuous manner. Video cameras and private security guards monitor compliance with the rules of the house. Shopping centers are not public spaces, but semi-public at best. Unwelcome visitors cannot participate in the "*urban life*" of the artificial spaces. This weighs all the more heavily the more functions beyond mere shopping are integrated into the shopping centers. In some cities, such as Minneapolis and St. Paul

whom spaces may be used, while the socially weak are at best tolerated or even excluded altogether (Harvey 2003, p. 32). There has been a public debate for years on the question of who owns the city and whether it is morally justifiable to systematically ban certain groups of the population from public space or certain neighborhoods (Davis 1992a, 2001; Harvey 2003; Mitchell 2003). Many locals, particularly in Los Angeles and New York, also resent the frequent closures of public space for filming (◘ Fig. 4.41). For cities, the temporary rental of public space is good business, but residents are denied access to parks and other locations for days at a time (Sorkin 2009, p. 114).

Individual transport has fundamentally changed the North American city within a few decades. More emphasis has been placed on moving and stationary traffic than on pedestrians. In many downtowns, parking garages and lots occupy more space than buildings, and freeways isolate downtowns from adjacent neighborhoods; neighborhoods are surrounded by streets that are nearly impossible to negotiate on foot, and neighborhoods lack sidewalks that might invite a stroll. Garage doors have replaced the front doors of private residences, and public buildings are no longer entered through prestigious entrances or even via open staircases, but by elevators that take visitors from underground garages directly to the floor they are aiming for. Outside the city centers, shopping centers, office parcs and free-standing shops, banks and office buildings lie like islands in a sea of parking spaces. People no longer use the cities as pedestrians, but at best nod to each other from their cars. Informal contact has become extremely rare (Gratz and Mintz 1998, pp. 33–34). On the traditional shopping streets of the inner cities, no one had been excluded. Here all ethnic groups and members of different socio-economic groups met without any restrictions.

in Minnesota, many downtown buildings are connected by *skyways*, i.e. enclosed pedestrian bridges on the first floor or higher. Visitors to downtowns can drive into a parking garage and walk dry to offices, shopping centers, restaurants, and other amenities. These downtowns are oriented almost exclusively to the interior; the street level is conceivably uninteresting because shops with attractive display windows are lacking (Crawford 1992, pp. 21–22; Hahn 1996b).

4.8.2 Inner City Squares

The restrictive access of shopping centers has been discussed and often criticized since the 1980s, but it is not limited to them. In almost every city center, there are squares and green spaces that are privately managed and controlled. In recent decades, many cities have become safer and cleaner, but at the same time, the diversity of urban life is now limited by the increased use of private security guards and cameras. In New York alone, there are some 300 such places that taxpayers helped to create or build, but which are not in the care of the public sector. These include the atriums of Trump Tower on Fifth Avenue and the adjacent Sony Plaza and IBM Plaza, as well as the Winter Garden of the Financial Center in Lower Manhattan, but also many small parks such as Bryant Park behind the New York Public Library and Tompkins Square Park on the Lower East Side that are not initially perceived as private (Kayden 2000) (Fig. 4.42). Located in downtown Manhattan at the intersection of 42nd St. and 6th Avenue, the 3.2-acre Bryant Park was established in 1871, but for several decades was used almost exclusively by drug dealers and junkies. Because the park was above street level and barely visible from the outside, it was ideally suited for a wide variety of criminal activity. The gen-

eral public felt uncomfortable and avoided the park as a place to stay. Through private initiative, the Bryant Park Corporation was formed in the 1980s and the park was closed and remodeled for several years. When Bryant Park reopened in 1992, it was visible from all sides and equipped with new recreational areas and seating. Private security guards now maintain law and order, and crime is a thing of the past. The square is an oasis of recreation in the middle of New York, but it is also one of the most controversial "*public*" spaces in the city. An expensive ice skating rink in winter and a Christmas market are among its many attractions. Especially the Fashion Week, which took place for many years in spring and autumn in Bryant Park, met with great criticism. In large tents, world-famous designers presented their collections exclusively to invited guests. During those weeks, the public had no access to Bryant Park, which had become a good business by renting out space, with revenues of up to about $7 million annually. Eventually, protest against the fashion show became so strong that the mega-event was moved to Damrosh Park near the Lincoln Center of the Performing Art for the first time in 2010 (Madden 2010; Vanderkam 2011). Bryant Park is far more attractive today than it was in the 1970s. Mothers with children, shoppers, office workers, tourists, and homeless people of all ethnicities collectively use the space to rest, meet friends, people-watch, play chess, or work on laptops. The square still attracts undesirables such as drinkers, homeless people, or even just people who are noticeably poorly dressed and obviously have no other place to be during the day. But these people are in the minority and are not perceived as a nuisance by other visitors (Gratz and Mintz 1998, pp. 38–42). BIDs, which are intended to use private funds to enhance selected areas in downtowns, are subject to similar scrutiny as Bryant Park or shopping centers. Creating clean and safe spaces is considered an important prerequisite for revitalization. This includes permanent surveillance by cameras, private security forces and municipal police, as well as the eviction of undesirables if necessary (Marquardt and Füller 2008, pp. 126–127).

4.8.3 Homeless People Without Civil Rights

Since there are only very limited unemployment benefits and welfare payments in the USA and destitute homeowners quickly lose their property if they do not pay the mortgages, the number of homeless people is far higher than in Germany (Fig. 4.43). Since the shelters are usually only available at night, the homeless wander through the cities during the day with their belongings, which they often have loaded on shopping trolleys. Many homeless people avoid the unattractive and

Fig. 4.42 Bryant Park in Manhattan

4

allegedly unsafe shelters altogether and sleep in parks, in doorways or under bridges, where they are of course not always welcome. Especially when large numbers of homeless people are concentrated in certain locations, a feeling of insecurity arises among residents. The violent eviction of Tompkins Square Park on New York's Lower East Side caused a particular stir. After several unsuccessful attempts to evict the park's "residents," 300 police in riot gear arrived at five in the morning of June 3, 1991, woke up the 200 or so homeless people living there, and evicted them. The tents and everything the homeless had left behind in the face of their hasty escape fell victim to bulldozers. While the eviction of Tompkins Square Park was particularly thuggish, it was not an isolated incident. New York was supposed to be cleaned up and rid of all undesirable elements in the 1990s. Following New York's lead, other conservative cities such as Miami and Atlanta, as well as more liberal places such as San Francisco and Seattle, soon implemented similar measures (Smith 2010, pp. 203–207). The repeated violent confrontations in People's Park in Berkeley, not far from San Francisco, caused a particular stir, even if the background here was initially different. There had already been tensions between residents and students of Berkeley University in 1969. When an unauthorized demonstration against the Vietnam War

took place in the park, the National Guard intervened and a large number of people were injured. Over the next two decades, the plaza became an (illegal) residence for homeless people, and riots broke out again in 1989 on the 20th anniversary of the riots. Subsequently, the debate over the toleration of homeless people was taken up in other locations such as San Francisco's Golden Gate Park (Mitchell 2003, pp. 118–160).

Homeless people are an extremely heterogeneous group and by no means correspond to the prejudice that they are all mentally ill or too lazy to work. The proportion of war veterans is strikingly high, but given the large number of home foreclosures as a result of the housing crisis in recent years, the number of homeless families has also increased. Single homeless people are predominantly male (67.5%), but 65% of all members of homeless families are female. Blacks are greatly overrepresented, while whites are significantly underrepresented (Adkins 2010, p. 219 f.). Many citizens and business owners feel disturbed by the presence of the often begging or not entirely clean-looking people at their front door or in their neighborhood. In recent years, a number of cities on the West Coast have adopted *sit-lie ordinances* that prohibit or restrict the use of sidewalks by homeless people. Authorities can enforce evictions in the event of resident complaints. These ordinances exist in Seattle, Portland, Palo Alto, and San Francisco, for example, where more recently there has been a resurgence of young people with alternative lifestyles traveling from place to place on the sidewalks in Haight-Ashbury (□ Fig. 4.44). The neighborhood, named for the intersection of Haight and Ashbury streets, was the center of the hippie and beatnik movements in the 1960s, but has since been largely *gentrified*. Whether the subculture's followers have a right to stay on the sidewalks is debatable. Some residents remember their own "*wild*" days and are tolerant, while others argue that they are

□ **Fig. 4.44** Alternative café in Haight-Ashbury

not "*real*" homeless people whose presence should not be tolerated. The problem worsened as more homeless people began living on sidewalks in other parts of the city as the number of evictions increased since 2008 (MacDonald 2010). Finally, in November 2010, San Francisco passed the *Civil Sidewalk Proposition*, a policy that allows authorities to crack down on people lying or sitting on the sidewalk between the hours of seven and 11 pm. At the same time, assistance programs must be set up to help those who are displaced.

4.8.4 Gated Communities

As early as the late eighteenth century, wealthy citizens of St. Louis and New York retreated to private streets or controlled-access estates, and in the first half of the twentieth century, members of the upper classes or Hollywood actors in search of privacy, security, or prestige lived behind gates and walls in private neighborhoods (◻ Fig. 4.45) (Blakely and Snyder 1999, p. 4). Since the 1960s and 1970s, *gated communities* have become a mass phenomenon. The driveways and entrance gates are guarded by security guards or video systems. In less exclusive *gated communities*, the gates are opened by the residents themselves or even stand open (◻ Fig. 4.46). Signs indicate that private property is located behind the gates and that unauthorized persons are not allowed to enter. Different rules apply inside *gated communities* than outside (Atkinson and Blandy 2006, p. viii). The exclusive *gated community* Desert Park in Phoenix MA is even radar monitored. In addition, there are other gated *communities* in the large area, which in turn are separated from the surrounding area by walls and gates (Frantz 2001, p. 18) (◻ Figs. 4.45 and 4.46).

◻ **Fig. 4.45** Luxury *gated community* in Scottsdale (Greater Phoenix)

◻ **Fig. 4.46** Simple *gated community* in Scottsdale (Greater Phoenix)

Since the transition from *gated* to *non-gated communities* is fluid, there is a lack of precise data on the number of private settlements. It is likely that more than three million people lived in about 20,000 *gated communities* as early as 1997 (Blakely and Snyder 1999, p. 7). Based on U.S. Census Bureau data, the number of housing units in *gated communities* is believed to have increased 53% from 2001 to 2009 to more than ten million (New York Times 2012b 3/29/2012). *Gated communities* are particularly prevalent in the South and Southwest of the country, where housing development began relatively late and was predominantly by private developers. Homogeneous *master planned communities were* created for clearly defined target groups. In addition to *retirement communities*, *lifestyle* or *golf communities*, in which the houses are built around a golf course, are particularly common. As in a club, clearly defined needs are served. In the traditional sense, club properties are neither private nor public; however, they may only be used by a clearly defined group. Users share the cost of club goods, making them less expensive and affordable for all (Manzi and Smith-Bowers 2006, p. 154). Very exclusive *gated communities* with large lots and houses, all costing several million US dollars, are contrasted with *gated communities* with comparatively inexpensive houses and small lots and little or no recreational amenities, as even the middle class increasingly retreats behind walls or gates. Developers suggest a sense of security in their marketing but do not advertise it, lest they be held liable in the event of crime behind the walls. Whether *gated communities* are safer than *non-gated communities* is debatable. The planned *gated communities* are mainly located in suburban areas and are supplemented in the cities by so-called *city perches*. These include social housing estates with high crime rates that have been subsequently separated from the surrounding area by a fence in order to better control access. However, it is also possible that residents of individual streets feel unsafe and prevent public access through gates that all residents have to pay for. Here, fear is often greater than danger (Blakely and Snyder 1999, pp. 99–103).

Hardly any other form of settlement is as controversial as *gated communities*, whose residents voluntarily submit to very restrictive rules, which often go far beyond those of normal *common interest developments* (CIDs), and 24 h surveillance. Visitors may only pass through the gates after consultation with the residents, and even visiting hours may be regulated. Because *gated communities* offer extensive services, use by non-paying guests is limited. It is possible to have overnight guests in one's home for only 14 days a year, and a fee must be paid to the *homeowner association* for each additional night. With such deep cuts in privacy, one wonders about the reasons for living in a *gated community*, which is also expensive as all security measures cost money. In luxury developments, the search for security and exclusivity may be the deciding factor, while in retirement communities, fiscal reasons may be paramount. However, taxes are also low in *non-gated retirement communities* because there are no schools to maintain. In addition, many residents of *gated communities* believe that home values rise faster here than in freehold subdivisions (Pouder and Dana Clark 2009, p. 217). Low (2010) interviewed residents of *gated communities* in the Texas city of San Antonio and in the New York City borough of Queens, where this type of housing development is, however, very rare. Most interviewees felt that crime had increased in their previous residential locations; accordingly, the search for safety was important in their decision to live in a gated neighborhood. They also wanted to live in homogeneous neighborhoods and hoped that properties in *gated communities* would be better protected from a drop in prices. In addition, interviewees believed that they had moved up the social ladder by moving into a *gated community*. Although they felt safer than in their previous freehold neighbourhoods, the fear of strangers had not diminished as the surrounding fences or walls were not seen as impassable and there were always service providers such as tradesmen or gardeners within the *gated communities* about whom too little was known. It seemed that moving into a *gated community* increased the fear of people living outside (Low 2010). Property value retention is also by no means guaranteed. In planned *golf communities*, if all the homes are not sold, the golf courses are too expensive to maintain. Without a golf course, the homes lose much of their value (Berger 2007, pp. 140–141).

Nowhere in the USA is segregation greater than in *gated communities*, which are inhabited almost exclusively by whites of the same income class with an identical lifestyle. People live as if on islands and no longer perceive the society around them. They isolate themselves and live only among their own kind. Social processes outside their own walls are excluded. Ritzer (2003, pp. 128–130) refers to *gated communities* as "*islands of the living dead.*" This criticism may be exaggerated, but numerous complaints can be found on the Internet that U.S. Census Bureau employees unlawfully visited *gated communities* as part of the surveys for the 2010 census, from the perspective of the residents. These people do indeed seem to believe they are outside (or above) society. Of course, another problem is that there are many people who cannot retreat behind walls because they are too poor. Still others have to drive long detours because thoroughfares are lacking.

Selected Examples

Contents

5.1 · Los Angeles: The New Prototype of the North American City?

105

5

5.1 Los Angeles: The New Prototype of the North American City?

Los Angeles' population increased nearly tenfold from only 11,200 to 102,000 between 1880 and 1900, and there was early speculation on building land (Jackson 1985, p. 178). While the cities of the East Coast and Midwest had their main period of development during the industrialization period, the rise of Los Angeles did not occur until the post-Fordist period. The population of Los Angeles has long been far more homogeneous than that of East Coast and Midwestern cities. As late as 1980, less than 15% of residents were foreign-born and less than 20% were non-white. It wasn't until the following decades that rapid change set in. Whites now make up just under half of the city's population, and more than 40% of Angelenos are foreign-born. At the same time, the proportion of Hispanics and Asians has increased rapidly (Erie and MacKenzie 2011, pp. 105–106, 121; ▶ www.census.gov). The legal and illegal influx of mainly poorly educated immigrants from Mexico and the Central American countries is on the one hand a major problem for the city, but on the other hand guarantees a large army of cheap labour.

The cities on the East Coast and in the Midwest were laid out in a comparatively compact manner, as their main period of growth took place before the invention of the automobile, while Los Angeles grew with individual traffic and was designed as a car-oriented city. In the surrounding areas, urban sprawl had begun in the late nineteenth century when petroleum was found in many locations and industrialized suburbs were created (Jackson 1985, p. 179). Recognizing early on that the close integration of high-emission industry with residential function was a major burden on the population, the City of Los Angeles passed the *Residence District Ordinance* in 1908, which provided for a strict separation of different uses and precluded any mix of uses. No industry or commerce of any kind was allowed on residential sites, and not even Chinese-run laundries or retail. Since the ordinance also prescribed a very low density of development, it was almost impossible to organize daily life without one's own vehicle. Los Angeles soon consisted mostly of single-family homes arranged around smaller centers connected first by rail and later by highways. In 1915, the San Fernando Valley was incorporated, located in a valley basin east of the Santa Monica Mountains, which rose to a good 500 m. Middle and upper class families were thus given the opportunity to move into neighborhoods with an upscale suburban character far from any industry. In 1930, 93% of all residences in Los Angeles were single-family homes, their percentage nearly double that of Chicago. The enclave of Beverly Hills, surrounded on all sides by the city of

Los Angeles, is a special feature. At the turn of the century, the Amalgamated Oil Company had purchased large tracts of land here in hopes of oil discoveries. When expectations failed to materialize, investors subdivided a 13 km^2 site north of Santa Monica Boulevard into spacious lots and founded the city of Beverly Hills here in 1914, which quickly became the preferred residential location of the rich and super-rich. At the same time, workers and the socially disadvantaged were concentrated in the neighborhoods between downtown Los Angeles and the ports to the southwest (Jackson 1985, p. 179; Kotkin 2001, p. 4).

Los Angeles County and the adjacent Orange County to the south together form the MA Los Angeles, also known as Greater Los Angeles. Nearly 13 million people now live here in an area of just under 10,600 km^2. Los Angeles County is subdivided into 88 independent municipalities such as the cities of Los Angeles, Santa Monica, Inglewood, Pasadena and Long Beach as well as another 140 dependent municipalities and is the most populous county in the USA with almost 10 million inhabitants. The smaller Orange County with the cities of Anaheim and Irvine is similarly fragmented. The Los Angeles *combined statistical area* (CSA) also includes Ventura, San Bernardino and Riverside counties, some of which are very sparsely populated. In the CSA, about 18 million people, or just under half of all Californians, live in 177 cities, about 30 of which have been newly designated since the late 1970s alone. With the new communities, homogeneous political units were often deliberately created, thus separating *us* from *them*. The region is additionally overlaid by a network of more than 1000 authorities for special tasks. The individual responsibilities can hardly be understood by the citizens and are therefore considered undemocratic. At the same time, the city region is hardly governable (Dear and Dahmann 2011, pp. 72–75; Judd 2011, p. 9; ▶ www.census.gov).

5.1.1 Chicago School

Since the three classic models of the *Chicago School* have been an integral part of every textbook on urban geography for decades (Carter 1980, pp. 205–213; Heineberg 2006, pp. 109–117), only the most important ideas will be summarized here. The models were developed in Chicago because the city grew faster than any other in the nineteenth century due to high immigrant numbers, and because the first chair of sociology in the United States had been established at the University of Chicago in 1892. The sociologists Park, Burgess and McKenzie understood urban development as a socio-ecological process and published the anthology *The City* in 1925,

5

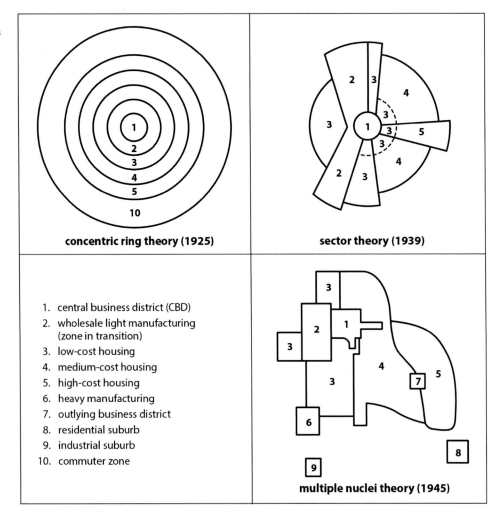

concentric ring theory (1925)

sector theory (1939)

1. central business district (CBD)
2. wholesale light manufacturing (zone in transition)
3. low-cost housing
4. medium-cost housing
5. high-cost housing
6. heavy manufacturing
7. outlying business district
8. residential suburb
9. industrial suburb
10. commuter zone

multiple nuclei theory (1945)

which is considered to be the first systematic attempt to explain the reasons and causes of urbanization (Park et al. 1925). The sociologists' interest focused on the possible dangers of rapid urban growth and the influence of European immigrants, who at the time came to Chicago mainly from Poland and other Eastern European countries, on existing structures. Empirical studies had established regularities in the spatial order of the city, which Burgess (1925, pp. 47–62) summarized in his model of concentric rings (◼ Fig. 5.1). Immigrants first moved into neighborhoods with poor, and therefore inexpensive, building stock near the center of the city, known in Chicago as the *Loop*. In a concentric ring around the *Loop*, the *zone of transition* emerged with the residential locations of the youngest generation of immigrants. Because immigrants from different countries of origin preferred *neighborhoods* where people from their homelands already lived, highly segregated neighborhoods formed. The *zone of transition* was also characterized by commercial and light industry and high crime rates. As assimilation increased, immigrants shifted to the adjacent concentric ring, where they now lived as second-generation immigrants in better homes and on larger lots, while a new generation of immigrants moved into the *zone of transition*. Over time, other concentric rings formed, into which people shifted as they moved up the social ladder. The most important finding of Burgess's research was that the center organized the periphery. The model of concentric rings created the basis for all further considerations of the spatial organization of the city.

Homer Hoyt studied at the University of Kansas and received his doctorate from the University of Chicago in 1933. His dissertation examined the development of land and rent prices in Chicago since the city's formation 100 years earlier (Hoyt 1933). Beginning in 1934, Hoyt served as an expert on real estate and land prices for the Federal Housing Administration (FHA) in Washington, D.C., for which he conducted detailed research in residential neighborhoods of 142 U.S. cities. He collected data on the level of rents in a total of 100,770 blocks and also compared information on residents, the condition of the houses and the length of tenancy. From this extensive material, Hoyt drew conclusions about the

5.1 · Los Angeles: The New Prototype of the North American City?

107

5

dynamics of urban development (Hoyt 1939). For the first time, the FHA was in possession of concrete data on the condition of individual neighborhoods. At the time, neighborhoods with ethnically mixed populations were considered unstable and uncreditworthy. Because these neighborhoods were circled with red paint on maps, so-called *redlining* became synonymous with neighborhoods for which loans were not approved (Beauregard 2006, p. 83). Hoyt has concluded that land prices did not decline uniformly from the city center towards the periphery, but that quite stable sectors of different land prices had formed. Land prices and rents were very low near industrial plants or transport routes. This was where the workers' housing was located, while the status-higher population lived in locations with fewer emissions. Based on his empirical research, Hoyt conceived a sectoral model of urban development. In 1945, geographers Chauncy Harris and Edward Ullmann, also at the University of Chicago, developed another model that focused on the use of each subarea of the city. They also attributed great importance to the city center, but noted that other centers had formed near the upscale residential neighborhoods, which in turn formed growth cores. This third model of the *Chicago School* is called the multi-core model (Harris and Ullman 1945).

For decades, the models of the *Chicago School* have received great attention, as they were the first and for a long time the only models for the spatial organization of the city. This is especially true for Burgess' model of concentric rings. However, it has often been overlooked that Wirth (1925) already pointed out that cities would change in the years to come in the face of new possibilities for communication and thus the relationship between center and periphery would also change (Dear 2005, p. 35). It was not until a new *School of Urbanism* (see below) began to form in Los Angeles in the late 1980s that the *Chicago School* models were increasingly questioned. It was criticised that Burgess' model was very general and dealt almost exclusively with the city center, gave no reasons for population displacement and did not take into account the locations of industry. Moreover, it only applied to very fast-growing cities in industrialized countries with a homogeneous surface (Hise 2002, p. 99; Erie and MacKenzie 2011, p. 111).

5.1.2 Los Angeles School of Urbanism

Los Angeles is undisputedly the most important location of the western film industry and the home of famous stars and starlets. Furthermore, the city has a rather bad image, as it embodies the urban sprawl par excellence and is known for extremely high air pollution (◘ Fig. 5.2). However, the Pacific metropolis is better

◘ **Fig. 5.2** View from the top floor of the Bonaventure Hotel over the sprawling city

than its reputation. Houston, Dallas, Atlanta, and San Jose have far lower population densities, and in fact Salt Lake City, Utah, which is located in a valley basin, has recently attracted attention for its far worse air quality. Probably due to its higher profile or the many universities in Los Angeles with dedicated geographers, a new *School of Urbanism* has been proclaimed in this city, with special attention to the fragmented settlement body, which contrasts with the old familiar *Chicago School* model (Dear 2002; Judd and Simpson 2011).

As early as 1959, Grey had described the most important characteristics of Los Angeles, such as low population density, decentralization of housing and services, ethnic polarization, the great importance of the automobile, and political fragmentation, and explained them by the late onset of growth (Grey 1959). Because Los Angeles differed in many ways from East Coast cities and even from San Francisco, he called it the new *urban prototype*. More accurately, Fogelson (1967, p. 2) wrote: "*Los Angeles succumbed to the disintegrative, though not altogether undesirable, forces of suburbanization and progressivism. And as a result it emerged by 1930 as the fragmented metropolis, the archetype, for better or worse, of the contemporary American metropolis.*" Elsewhere he compares Los Angeles with Chicago: "*It was not like Chicago – a typical concentrated metropolis – inhabited largely by impoverished and insecure European immigrants*" (Fogelson 1967, p. 144). The poor immigrants in Chicago had to live tightly packed together near workplaces, while the citizens of Los Angeles were relatively affluent even in the early stages of the city and knew the laws of the land market. However, it was not until some 20 years later that a loosely connected group of scholars discussed the postmodern development of the city and further developed existing theoretical approaches, following the spatial structures of the city of Los Angeles.

5

Much of the work on Los Angeles emerged in the late 1980s and 1990s, when the region was in recession due to the closure of military bases and the collapse of the aerospace industry in southern California. In addition, the race riots of the early 1990s were among the largest riots in U.S. history (Erie and MacKenzie 2011, p. 126). However, the approach of the *Los Angeles School of Urbanism* is controversial, and at the same time, a debate has erupted around the *Chicago School of Urbanism* and its evolution, now referred to as the *New Chicago School* or *Chicago Not-Yet-a-School of Urban Politics* (Clark 2011, p. 221). Since the publications on the aforementioned schools now take up entire shelf walls, only the most important statements can be reproduced here. This also applies to the extensive criticism of the individual schools.

Los Angeles is considered a postmodern city par excellence (Caves 2005, p. 93). There are many different definitions of the term "*postmodern*", but it always stands for a radical break with everything that has gone before, i.e. for a discontinuity between past and present (Dear and Dahmann 2011, p. 68). The West Coast metropolis differs in many ways from Chicago and other cities of the American East Coast and Midwest. This is true of the low population density, sprawl and fragmentation, the large number of secondary downtowns and the associated decentralized construction of the city, the now exclusively privately organized construction of new neighborhoods, the great ethnic diversity of the population and the numerous ethnic neighborhoods, as well as the fortress-like mega-complexes and numerous man-made theme worlds. Approaching Los Angeles by air, one not only sees an almost endless urban landscape, but also has trouble clearly identifying the old downtown, as there are several other centers, such as that of Hollywood or Mid-Wilshire, that stand out from the surrounding area with similar numbers of high-rise buildings as the historic center of Los Angeles. Garreau (1991, pp. 431–432) had identified 16 fully developed *edge cities* and eight others that were on their way to becoming *edge cities* in the Los Angeles MA. Even though Lang (2003, p. 136) could clearly identify only six *edge cities* based on his empirical research, the large number of more or less equal centers in the region becomes clear. Another characteristic of the region is neighborhoods that are organized by private interest groups and constitute *privatopias*. Greater Los Angeles has more than 11,000 CIDs with nearly 1.7 million households. The proportion of privately managed neighborhoods will continue to increase in the coming years, as currently all neighborhoods in planning or under construction, without exception, are organized as CIDs (Community Association Institute 2013). Few other cities in the U.S. have as much ethnic diversity as Los Angeles. Almost 50% of the population are Hispanics and just under 29% are non-Hispanic whites. A good 11% are of Asian origin with a large proportion of Koreans and another just under 10% of the population belong to the black minority. More Mexicans, Salvadorans, Bulgarians, Hungarians, Israelis, Iranians, Armenians, Ethiopians, Filipinos, Thais and Koreans live in Los Angeles than any other place in the world outside their home country. In addition, more Japanese and more members of Native American original populations live in Los Angeles than in any other place in the United States (▶ www.census.gov). Probably more than 600,000 people also belong to the Jewish faith; after New York, Los Angeles thus has the second largest Jewish population in the United States. The city is composed of a large number of ethnic enclaves with their own identities such as Chinatown, Koreatown, Thai Town, Filipinotown, Little Tokyo, Little India, Little Armenia, Little Ethiopia, and Little Persia. The individual ethnic groups are under a lot of pressure to adapt and maintain their own identities. The cultural dynamics and the coexistence of the many ethnic groups, which Dear and Flusty (1998, p. 54) call the "*culture of heteropolis*", are reflected in heterogeneous architecture, but also in social inequality, polarization and occasional social unrest. Los Angeles continues to be typified by the large number of theme parks and artificial worlds that simulate the city but have no place-typical reference themselves. Soja (1992) refers to Orange County as *simulacra* because it copies the main features of a city, but is itself a fake (see also Baudrillard 1994). These "*urban*" theme parks are complemented by numerous amusement parks such as Disneyland in Anaheim, Universal Film Studios in Hollywood with the adjacent Universal CityWalk with restaurants, nightclubs, and experiential shops. Irvine Spectrum Center in Orange County is an oriental city style shopping center with a wide range of dining options. *Urban entertainment center* themes are used to enhance marketing and provide the customer with the greatest possible experience (Hahn 2001). In Los Angeles, not only are the amusement parks and shopping centers not accessible outside of opening hours, and in the *gated communities* strangers are completely denied access; in addition, there are many repellent places where a part of the population is obviously not wanted. This is particularly true of parts of downtown Los Angeles, whose decline in importance was initiated in the 1940s. By the 1960s, most of the old bank buildings were vacant. Since the 1970s, numerous office buildings and hotels such as the Bonaventure Hotel, the World Trade Center and the Arco Center, but also a new *entertainment district* with the Disney Concert Hall designed by Gehry, have been built on Bunker Hill with financial support from public funds (◼ Fig. 5.3). Davis already deplored the fortress-like and forbid-

5.1 · Los Angeles: The New Prototype of the North American City?

109

5

Fig. 5.3 Bunker Hill in Los Angeles

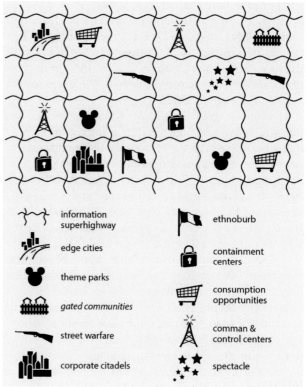

information
superhighway

ethnoburb

edge cities

containment
centers

theme parks

consumption
opportunities

gated communities

comman &
control centers

street warfare

spectacle

corporate citadels

Fig. 5.4 *Los Angeles School of Urbanism.* (Adapted from Dear 2005, Fig. 4)

ding character of Bunker Hill in the early 1990s and criticized its self-sufficiency: "*The new financial district is best conceived as a single, self-referential hyperstructure, a Miesian skyscape of fantastic proportions*" (Davis 1992b, p. 157). *Miesian* refers to the forbidding and predominantly black-clad office buildings designed by the German-born architect Mies van der Rohe. Between the citadels of the large corporations are attractive private plazas that are inaccessible in the evening and at night to keep out the unwanted homeless. Davis (1992b) criticizes this state of affairs in the strongest terms, using vocabulary such as *forbidden city*, *fortress*, or *militarization of urban space*. In the past 20 years, and especially since the terrorist attacks in 2001, surveillance of public space has been further expanded.

The locations of individual uses are chosen seemingly aimlessly in Los Angeles. Because all land is connected by information highways, complementary uses need not be adjacent. Like the game of keno, capital blindly chooses certain sites and ignores adjacent sites. Quasi-randomly, a collage of adjacent developed and undeveloped land emerges. The isolated land can be developed over time, but need not be, as the fragmented urban landscape expands unchecked in every direction (Dear and Dahmann 2011, p. 70). Based on their observations in Los Angeles, Dear and Flusty (1998, p. 66) developed the *Los Angeles School of Urbanism* model (■ Fig. 5.4), which negates the *Chicago School's* concentric rings model. While Burgess assumed that the center organizes the hinterland, in the new model the hinterland determines "*what remains of the center*" (Dear 2005, p. 35). Postmodern *urbanism* reverses the logic of modernity, according to which cities expand from the center towards the periphery and development is controlled by the center. According to postmodernism, the center no longer dominates the periphery, but rather the periphery organises the center, if the latter still exists at

all. Suburbanization in the classical sense, i.e., the relocation of population, industry, and services from the center to the surrounding areas, no longer exists in Los Angeles. Even the term *sprawl is* outdated, according to supporters of the *L.A. School of Urbanism.* In southern California, urban formations without a center have been springing up for some time, some of which are added to later for aesthetic and marketing reasons, or just to encourage consumption. However, these urban centers are only meaningless externalities of postmodern urban development (Dear and Dahmann 2011, pp. 68–76).

The representatives of the *L.A. School of Urbanism* are of the opinion that the city of Los Angeles is the prototype for the development of urban spaces not only in the USA, but worldwide (Erie and MacKenzie 2011, p. 107). They argue that numerous studies have now empirically demonstrated that other American cities have developed like Los Angeles in more recent times. Scott (2002) has documented for several cities in California that industry has chosen different locations than in old industrialized cities, thus influencing spatial development. Similarly, in New York, Chicago, and Washington, D.C., populations have become more ethnically heterogeneous at the same time that the dichotomy of whites and blacks has dissolved (Myers 2002). *Edge cities* now exist in many large cities, and private lifestyles are ubiquitous in the United States. Moreover, post-Fordism and the network society have changed the demands on production sites everywhere. Dear and Dahmann (2011, pp. 66–77) therefore believe that Los Angeles is a model for the future development of all American cities. Even Chicago, once the most important representative of modernity, is now developing as shown in the *L. A. School* model.

The *Los Angeles School of Urbanism* model has been widely criticised, although it is important to bear in mind that some of the critics take different scientific approaches to explaining urban development processes than the proponents of the new city model. Abu-Lughod (2011, p. 21) justifies the differences and similarities between the three cities of New York, Chicago and Los Angeles from a historical perspective and stands by the fact that she is not a theorist. Beauregard (2011, pp. 195–199), who is far more theoretical than Abu-Lughod, also emphasizes that spatial and historical developments should not be neglected in the search for similarities and differences between cities. Overall, he is critical of the formation of theories based on developments in a single city. Bridges (2011, pp. 97–98) faults the *L. A. School of Urbanism* for examining the spatial and economic dimensions of the city but neglecting the people living in the city. Erie and MacKenzie (2011, p. 118) criticize the economic determinism of the *L. A. School of Urbanism*, but this also underlay the concentric rings model. While the latter represented urbanization brought about by industrialization, the *L. A. School of Urbanism* depicts development brought about by postmodernism. Clark (2011), on the other hand, finds fault with the accounts of the fragmentation of Los Angeles by Mike Davis and Michael Dear as a whole, as they are based only on the contrasts of the affluent (*haves*) and the less affluent or poor (*have-nots*). *The L. A. School of Urbanism* model is based on the idea that the downtown area is of little importance, however, major investments even in downtown Los Angeles have initiated a significant resurgence in recent times (Spirou 2011, p. 278). At the turn of the millennium, the $375 million Staples Center and the $274 million Walt Disney Concert Hall opened. At the same time, the Cathedral of Our Lady of Los Angeles and Los Angeles City Hall underwent extensive renovations. Due to the positive population development, a supermarket was opened again in downtown Los Angeles in 2007 after several decades.

5.1.3 Outlook

As a counter-position to the L.A. School of Urbanism, a *New York School* has developed that denies the dissolution of the urban center and doubts that the new West Coast model (*sprawled, centerless and fragmented*) represents a new prototype of the American city. While there is no cohesive *New York School*, certain developments are indisputable among East Coast geographers. These include the great importance of the center and the recognition that growth of suburban space does not limit its vitality. It is also acknowledged that policy influences urban development and that the allocation of public resources is important (Mollenkopf 2011, p. 171). According to, the *New York School*, which focuses on the potential and importance of central city areas and Manhattan in particular, has its origins in the work of Jane Jacobs and her contemporaries in the 1950s. The advocates of this school, they say, are determined to defend and improve urban life. They are convinced that life in the city should be preferred to life in suburban areas and that in core cities the wealthy, the middle class and the poor can live side by side. In addition to Jane Jacobs, the architect Robert Stern, who developed early thoughts on the revitalization of Times Square, the sociologists Richard Sennett and Sharon Zukin, and the urban planner William Whyte had been important exponents of the *New York School* (though they did not describe themselves as such) because of their awareness of the value and quality of life in the city. Zukin (1982) had provided an early work on the *gentrification* of New York with her analysis on the conversion of previously commercial lofts into apartments (Halle and Beveridge 2011, pp. 138–139). It should not be overlooked that Burgess's 1925 model does not apply to the city of Chicago today either; however, the *L. A. School* model cannot explain Chicago's structure today either. Although Chicago is now also a highly fragmented city, there is still a very dominant center. Ultimately, it is unnecessary to argue about the existence of a *Chicago, Los Angeles*, or *New York School*; it is more important to highlight certain characteristics that shape or distinguish the cities (Mollenkopf 2011, p. 182).

5.2 Lower Manhattan After 11 September 2001

When people think of the USA, they often think of New York or, more precisely, Manhattan, although the east coast metropolis is not typical of the US city in many respects. New York is comparatively old, very compact, and the private car plays only a minor role. New York has experienced many crises, but was hit to the core by the terrorist attacks on September 9, 2001, which brought down the World Trade Center and several neighboring buildings, and at times even the future of the city was called into question. During and after the terrorist attacks in September 2011 and during and

Fig. 5.5 World Trade Center in New York. (Adapted from Pries 2002, Fig. 2)

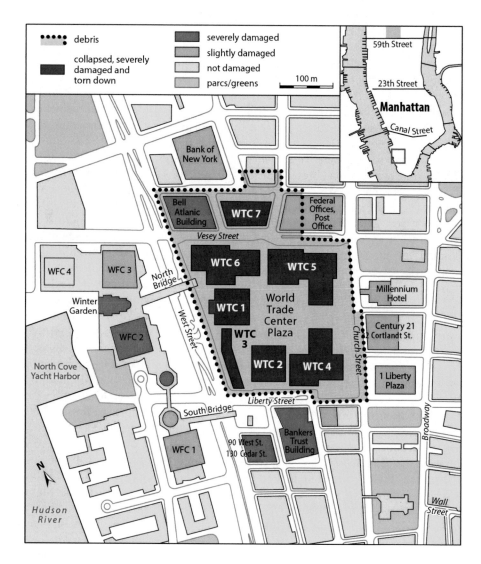

after Hurricane Sandy in the fall of 2012, which caused extensive damage to coastal areas, New Yorkers were repeatedly described on radio and television as tough or resilient. New Yorkers fight for their city and do not allow themselves to be driven away by terrorist attacks or forces of nature (observations by the author, Hahn 2004).

Manhattan south of Central Park is divided into three parts. Midtown is located between 59th St. and 23rd St., and is joined by Midtown South to the south as far as Canal St. The blocks at the southern tip of Manhattan are referred to as Lower Manhattan or Downtown (Fig. 5.5). Construction of a World Trade Center (WTC) in New York had been discussed as early as the 1940s, but it was not until the 1960s that construction began on the twin towers visible from afar on the banks of the Hudson River in Lower Manhattan. The complex of seven buildings was commissioned by the Port Authority of New York and New

Jersey and designed by architects Minoru Yamasaki and Emery Roth & Sons. Completed in 1973, the two towers, WTC 1 and WTC 2, each 417 m tall, had 110 stories. The other five buildings in the complex, WTC 3 to WTC 7, were much smaller and were opened by 1987. In addition to offices, the World Trade Center included a shopping mall, a hotel, and the underground station of the PATH trains to New Jersey. The excavation for the new building complex was piled up on the Hudson River south of the WTC. This is where Battery Park City, a high-rise housing development with upscale apartments, was built. In addition, the Financial Center with a further 0.8 million m^2 of office space was opened on freshly excavated land between the WTC and the Hudson River. Although the Twin Towers were not the tallest buildings in the USA and were towered over by the Sears Tower in Chicago by around 100 m, the World Trade Center was regarded as a symbol of capitalist America (Pries 2002, p. 61, 2005, pp. 8–9).

5

5.2.1 **Destruction**

The two Boeing 767s that flew into the World Trade Center on September 11, 2001, brought down the Twin Towers. The other five buildings in the complex collapsed under the weight of debris from WTC 1 and WTC 2 or were so badly damaged that they had to be demolished. Another 23 buildings in the World Trade Center neighborhood were badly damaged. A total of 2752 people were killed. More than 20,000 residents of Lower Manhattan had to temporarily leave their homes. The train station located under the WTC was also destroyed, and several subway lines and train connections to New Jersey had been severed. In addition, hundreds of shops and restaurants have had to close temporarily or permanently (Alliance for Downtown New York 2010, p. 4). In the seven WTC buildings, 1.23 million sq. ft. of office space, or 12.5% of Lower Manhattan's office space, was destroyed, and another nearly 2 million sq. ft. in 23 adjacent buildings was damaged. In total, 60% of Lower Manhattan's Class A office space was unusable. Relocation of jobs was inevitable. Fifty thousand financial sector workers lost their jobs. An advantage was that, as a result of the recession that had set in after the turn of the millennium, many office properties in Manhattan were vacant at the time of the attacks. Almost a third of the destroyed space could therefore be rented immediately at other locations in Manhattan. Offices for around 19,000 employees were rented in suburban areas, with Hoboken in New Jersey, located on the other side of the Hudson River, being preferred (Gong and Keenan 2012, p. 373; Pries 2005, pp. 4–6).

5.2.2 **Reconstruction**

The terrorist attacks and the scale of destruction deeply affected Americans, but left little doubt that Lower Manhattan would be rebuilt. One of the few exceptions was Mike Davis (2001, p. 5), who argued that the attacks unquestionably spelled the end of the revitalization of downtown New York and other inner cities. He predicted that job decentralization would proceed apace and that the high-rise would soon be a thing of the past. Davis feared that the flight of the two Boeing planes into the World Trade Center would have as destructive an effect on the future of the skyscraper as the impact of an asteroid once had on the dinosaurs. It soon became apparent that this pessimistic assessment was wrong. New Yorkers did not leave the city in large numbers, and high-rise construction did not stop. From 2002 to 2011, 39 buildings taller than 150 m were completed in New York (CTBUH 2011, p. 55). Most of the jobs

that had been forcibly outsourced to other locations were moved back to the Downtown as soon as possible. The great prestige and feel-good benefits were valued more highly than the fear of further attacks (Gong and Keenan 2012). Manhattan remains a popular location for service businesses and attractive to career-minded college graduates. In 2010, weekly wages in Manhattan averaged $2404 and were 170% higher than the national average and as much as 45% higher than Silicon Valley (Glaeser 2011a, p. 5).

Surprisingly, Lower Manhattan has developed particularly positively in recent years. The upswing is not due to the increased media attention after the attacks, but had already begun in the 1990s. At the southern tip of Manhattan, which is characterized by a layout laid out by the Dutch with relatively narrow and sometimes winding streets, the city's first high-rises had been built since the end of the nineteenth century and were mainly used by financial service providers. The narrow development soon hardly allowed for the construction of new buildings. Since the 1930s, and increasingly since the 1950s, high-rise buildings were erected primarily in Midtown Manhattan along 5th Avenue or 6th Avenue. Lower Manhattan increasingly showed signs of decay similar to those of other American cities. Many of the old office buildings no longer met the new technical requirements and could only be leased on poor terms. In 1993, 26.4% of office space was vacant, a figure surpassed only by Dallas at the time. Problematically, the southern tip of Manhattan was largely deserted after offices closed and on weekends. In 1995, property owners and business owners formed the *business improvement district* Alliance for Downtown New York with the goal of improving the quality of life for residents, employees, and visitors. This was to be accomplished by providing social services, maintaining cleanliness and safety, planting and maintaining green spaces, as well as better marketing. At the same time, the *Manhattan Revitalization Plan* was adopted for a period ending in 2001, which mainly promoted the location of new businesses by means of tax reductions. The plan supported the conversion of old office buildings into residential buildings. Since many of the predominantly older buildings on Wall Street no longer met the requirements of modern offices, some were converted into apartment buildings. The number of residents in Downtown was increasing. In addition, several hotels were under construction or in planning. The turnaround from a monofunctional district to a multifunctional use was clearly visible when the World Trade Center was destroyed in 2001 (Gong and Keenan 2012; Pries 2001, 2005, p. 8; ▶ www.downtowny.com).

The American government and the State of New York have generously supported the reconstruction of Lower Manhattan financially. In 2002, an architectural competition was announced for the future development of the approximately 6.5 ha World Trade Center site, which has been referred to as *Ground Zero* since the terrorist attacks. As early as February 2003, the winner was announced as Daniel Libeskind, who was from then on responsible for the implementation of the overall concept. However, the individual buildings were constructed by different architects such as David Child, who designed the tallest building on the site and whose foundation stone was laid in 2004. Instead of the twin towers, only one building will be built, which will be the tallest skyscraper in the U.S. at 541 m when it is scheduled for completion in 2014. Four other office buildings will also be constructed on the site. Until 2009, it was planned to name the new tower "*Freedom Tower*"; however, the official name is now "*One World Trade Center*" (◘ Fig. 5.6). Almost all floors are intended for office use. As in the previous buildings, shops and restaurants will be located in the basement, and an observation deck will be built on the top floor. When completed, One World Trade Center will be as iconic a building as the former World Trade Center, and naysayers fear, an equally desirable target for terrorists. However, the new building reportedly cannot collapse like the Twin Towers, and security is said to be far better. Since some

◘ **Fig. 5.7** National September 11 Memorial

of the office space has long been leased, there seems to be little reluctance to move into the tallest building in the US. As far as is known, Condé Nast will be one of the biggest tenants. A good 90,000 m² in One World Trade Center was reserved for the publishing house back in 2011, but leasing has been rather stagnant since then. The areas where the Twin Towers stood have been left out of the development. Here, on the tenth anniversary of the attacks, the National September 11 Memorial was opened in honor of the victims (◘ Fig. 5.7). Two ten-meter-deep square granite basins were constructed on the footprints of the destroyed towers, with water flowing continuously into them from all four sides. An associated museum is scheduled to open in spring 2014 (Alliance for Downtown New York 2011; Lewis and Holt 2011; ► www.911memorial.org).

Given the quick decision to rebuild *Ground Zero*, the private sector has also invested in Lower Manhattan. Before the attacks, 325,000 people worked in the Downtown; in 2011, the number was back up to 309,000. At the latest, the completion of One World Trade Center will surpass the 2001 figure. More office buildings are under construction or in the planning stages. The largest employers in Lower Manhattan are still in the *FIRE* sector (Finance, Insurance, Real Estate). The Bank of New York Mellon Corp., Deutsche Bank and Merrill Lynch & Co. alone each have a good 10,000 employees. Outside the *FIRE* sector, major employers include the City and State of New York, the U.S. government, and the Metropolitan Transportation Authority. From 2005 to 2011, 307 businesses moved to or were established in Lower Manhattan. At the same time, a diversification of the economy has begun. The media sector, previously hard to find on the southern tip of Manhattan, was represented by more than 60 companies in 2011, including Broadcast Music, American Media, Inc. and New York Daily News. As demand for office space has increased,

◘ **Fig. 5.6** One World Trade Center

5

rents in Downtown have picked up, though they are still lower than in Midtown Manhattan (Alliance for Downtown New York 2011, pp. 7–8, 13).

Housing construction in Lower Manhattan was supported with public funds after 2001. A peak was reached in 2008, when 3300 housing units came on the market. The number of downtown residents nearly doubled in a decade. Ten years after the terrorist attacks, 56,000 people were living in 312 buildings with just over 28,000 apartments, most of which were in the luxury segment. By 2014, a further 2546 apartments will be ready for occupation. The vacancy rate is extremely low at 0.7%, as the apartments are in great demand. The average rent for an apartment is nearly $4000 per month (as of 2011), nearly 20% more than the rest of Manhattan. The Beekman Tower, designed by star architect Frank Gehry, attracted particular attention when it was occupied near the East River in early 2011. With 76 floors and a height of 267 m, the striking structure is the tallest residential building in the USA. Lower Manhattan residents are better educated than average, and household incomes are roughly double those of Manhattan as a whole. Although Downtown is dominated by highrise office buildings and open space is rare, children live in a quarter of households. In a survey, those living in Lower Manhattan said they appreciated the high quality of life, high-quality amenities, excellent access to mass transit, good schools, and high safety standards. About 30% of respondents were able to walk to their jobs. In addition, the parks, waterfront, and good recreational opportunities were rated positively (Alliance for Lower Manhattan 2010, pp. 15–17).

Before 2001, tourists were hardly interested in a hotel in Lower Manhattan. This has changed as the economy has diversified and the number of residents has increased. In addition, *Ground Zero* became an attraction that benefited other landmarks such as Federal Hall, the New York Stock Exchange, St. Paul's Chapel, and Trinity Church. From 2005 to 2010, the number of tourists to Downtown has steadily increased from four to nine million. In the future, it is estimated that another three to four million tourists will view the September 11 Memorial, which opened in 2011. The number of hotels tripled from 6 to 18 from 2001 to 2011, and the number of hotel rooms increased by 78% to 4092. In 2011, another seven hotels with nearly 1000 hotel rooms were under construction or approved. The hotels are predominantly in the luxury segment and 55% are booked by business travellers. Although the average nightly rate was $283 in 2011, 23% higher than the rest of Manhattan, the occupancy rate was very good at 80–85%. In parallel with the number of residents and visitors, the number of retail-

ers and restaurants in Lower Manhattan has increased sharply in recent years.

Part of Lower Manhattan's success is due to its outstanding access to mass transit. Fourteen subway lines connect the borough to the rest of New York, and the underground PATH train, whose station at the World Trade Center had to be completely rebuilt after 9/11, provides a quick ride to New Jersey. The subway and new stations operate more efficiently today than before the attacks. In addition, the southern tip of Manhattan can be reached by 33 bus lines and several ferries. A heliport is available for those in a hurry. Good public transportation means that Downtown residents can reach their workplace in an average of 22 min, 17 min faster than other New Yorkers.

5.2.3 Outlook

In October 2012, residents, employees and hotel guests of the Downtown had to learn painfully that danger is not only threatened by terrorists (◻ Fig. 5.8). Hurricane "*Sandy*" destroyed large parts of the mid-Atlantic coast. In New York, parts of Long Island and Lower Manhattan were hit hard. The power was out for days, water ran into subway shafts and penetrated shops, restaurants and the lobbies of office and residential buildings, especially on the banks of the East River. Complete chaos reigned for a few days, but accustomed to dealing with disasters, New Yorkers quickly cleaned up the damage and normal life soon returned. In view of climate change, it cannot be ruled out that such natural disasters will become more frequent in the future.

◻ **Fig. 5.8** Near Pier 17 in Manhattan, Hurricane "Sandy" destroyed businesses

5.3 Chicago: From Industrial to Consumer City

Chicago (◘ Fig. 5.9) has always had a far more diversified economy than monostructured Detroit, and the conditions for successful structural change have been comparatively good. Today, more than 90% of all employees in the city work in services and less than 10% in industry. Banks, insurance companies, the futures exchanges, and numerous corporate headquarters are important employers, as are the universities, the hotel and restaurant industry, and retail trade (▶ www.census.gov; ▶ www.fortunemagazine.com; Hahn 2004). Chicago is a highly segregated city. While the south and parts of the west of the city are characterized by severe poverty, the wealthy citizens are concentrated in the north. Throughout Chicago's short history, violence and rioting have occurred, and as recently as the 1980s, the ethnically polarized and violence-prone city was referred to as "*Beirut by the Lake*" (Bennett 2010, pp. 5–7). In the ensuing decades, crime rates have dropped; however, as the city's deficit neighborhoods are increasingly terrorized by rival gangs, the murder rate has risen again in more recent times and is currently far higher than in New York (New York Times 6/25/2012c). Those living on the North Side and the tourists who most rarely or never travel to the ghettos on the South Side do not perceive the many problems.

◘ **Fig. 5.10** View of the *Loop* with Willis Tower in the background

Downtown has become a showcase for a modern city and now attracts suburbanites and national and international visitors in large numbers (◘ Fig. 5.10). Downtown has undergone a process of *gentrification* and functional transformation unlike any other American city. After decades of decay, high-rise apartment buildings are sprouting up like asparagus, old office buildings are being converted into housing, hotel towers and shopping centers are being built, international designers are opening flagship stores, parks and green spaces are being created, and attractive recreational facilities are opening. An inner city that was long considered dangerous and had the worst image imaginable has developed into a lively center where people enjoy sitting in street cafés or pursuing other pleasures until late at night in the summer.

5.3.1 Urban Redevelopment

Chicago has always been characterized by successful entrepreneurs, but their success was due to the exploitation of the working class. Probably no other city in America was shaped more by capitalism than the metropolis on Lake Michigan. Chicago was always in competition with New York. An entrepreneur who had made the grade in Chicago did not go to New York, but campaigned for Chicago to be bigger, prettier, and better than its East Coast competitor. Chicago never lacked self-confidence: "*North Michigan Avenue is the Fifth Avenue of the Middle West; and already it looks forward to the day when Fifth Avenue will be the North Michigan Avenue of the East*" (Zorbaugh 1929, p. 7). Wealthy citizens always supported the city ideally and financially, and the Chicago Commercial Club already sponsored the *Plan of Chicago* (Burnham Plan) by Daniel Burnham and Edward Bennett, published in 1909 (Burnham and Bennett 1909).

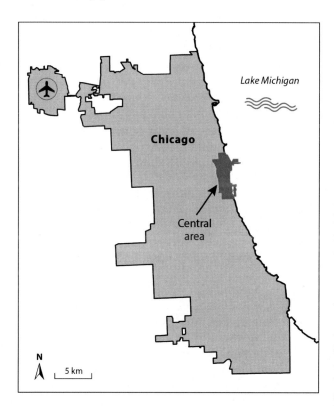

Lake Michigan

Chicago

Central area

N

5 km

◘ **Fig. 5.9** Chicago

5

Mayors, by their very nature, still had far more influence than engaged citizens. Mayor Richard J. Daley (1955–1976) succeeded in transforming Chicago into a post-industrial city. The construction or expansion of major infrastructure projects such as O'Hare Airport, the McCormick Place convention center, the University of Illinois campus, and the subway and urban freeway networks were particularly important to him. The city coordinated land use and showed great success in attracting public and private funding. In 1958, Daley founded the Department of City Planning, which adopted the *Development Plan for the Central Area of Chicago* that same year. The long-term goal was to transform the city center into a central service hub for the entire region. New parks and recreational facilities on the shores of Lake Michigan were to enhance the quality of life in the central areas of the city, and modern and futuristic architecture was to follow earlier architectural achievements. Housing for 50,000 people of all income levels was to be built in the *Loop* alone (Bennett 2010, pp. 5–7, 39; Clark et al. 2002, pp. 503–508; Newman and Thornley 2005, p. 105). The 1973 *Chicago 21: A Plan for the Central Area Communities* plan was an update of the 1958 plan, but sparked intense controversy because it placed a very high emphasis on building high-end housing for high-income buyers or renters in the downtown area. The 1958 and 1973 plans both harkened back to the 1909 Burnham Plan. The world's best architects were to erect modern buildings, open space was to break up the downtown, and the lakefront was to be enhanced. The intent was to prepare Chicago for a great future as a national, if not global, service center leader (Bennett 2010, pp. 38–43).

The 1983 *Chicago Central Area Plan: A Plan for the Heart of the City* had similar goals. With the expansion of the McCormick Place convention center south of the city center, the construction of a museum campus with a planetarium, aquarium, and the Field Museum of Natural History, and the upgrading and beautification of major streets, the city center was to become more attractive (City of Chicago 2003, p. 6). The *1984 Chicago Development Plan* was adopted by Harold Washington (1983–1987), the city's first and so far only black mayor, and under the guiding principle of "*Chicago works together*" no longer pursued the goal of securing Chicago's position as a major service center and world city. Rather, the consequences of deindustrialization were to be eliminated by creating new jobs, especially for the disadvantaged black population, and by building social housing. After Washington's sudden death in 1987 and two interim mayors, Richard M. Daley, a son of Richard J. Daley, became mayor in 1989. Daley held the office until 2011, when he resigned due to age. He ran the city like a business, focusing on the redevelopment of the Central Area. This includes the historic center of the

■ **Fig. 5.11**   View from Willis Tower towards the North Side

city south of the Chicago River, referred to as the *Loop*, the Near North Side north of the river (■ Fig. 5.11), the Near South Side south of the *Loop*, and the Near West Side west of the Chicago River. Mayor Richard M. Daley, like his father several decades earlier, championed the implementation of major infrastructure projects (Bennett 2010, pp. 6, 43–45). Although there had been plans for downtown revitalization since the 1950s, by the late 1980s it was still very unattractive and characterized by signs of decay and vacancy. Investment had been limited mainly to the public sector, such as the construction of the State of Illinois Center (now the James R. Thomson Center) in the *Loop*, which opened in 1985. Even the construction of the Sears Tower (today Willis Tower) in the early 1970s, which at 442 m was the tallest building in the USA for decades, had not been able to initiate an upswing. However, the preconditions for the creation of an attractive city center were not bad and had already been laid by the Burnham Plan of 1909. Grant Park, 3.2 km long and about 800 m wide, had prevented the development of the lakeshore east of the *Loop*. Moreover, since the best architects in the United States had worked in Chicago, the city had many interesting buildings (Cremin 1998, pp. 22–26; Judd 2011, p. 16; Larson and Pridmore 1993).

Richard M. Daley has finally succeeded in implementing the long-standing plans for the reconstruction of the Central Area (■ Fig. 5.12). This has been done largely through public-private partnerships as well as private donations. One of the first projects to be realized was the redevelopment of Navy Pier, opened in 1916, which had been used first as a dock for cargo ships, then for amusements of all kinds, by the Navy during World War II, and then as a campus by the newly established University of Illinois until 1965. In the mid-1990s, at a cost of more than $200 million, the roughly one-mile-long pier was redeveloped into a major recreational facil-

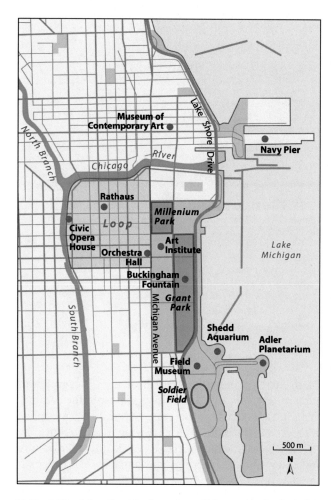

Fig. 5.12 Attractions in downtown Chicago. (Author's design based on plans for tourists)

also been expanded and redeveloped. In 1948, Meigs Field airport had opened on an artificial island in Lake Michigan, which was closed in 2003 to create a park on the approximately 1.2 km long island. The aforementioned projects were very significant for the development of the southern downtown and adjacent areas (Spirou 2006, p. 297, 2011, pp. 284–285). Impressive is the transformation of the Chicago River waterfront, which not long ago had served as a transport canal and the site of the city's major industries. With deindustrialization, numerous areas along the river had been abandoned and put to new uses. The waterfront is now used for leisure and recreation, and a waterfront promenade with cafés and attractive terraces has sprung up along the riverbank (City of Chicago 2003, p. 11).

5.3.2 Millennium Park

The city of Chicago received a new face, however, especially in 2004 with the opening of Millennium Park north of the Chicago Art Institute on a long-abandoned railway site between Lake Michigan and the northern *Loop*. The park cost around 500 million US dollars to build, around half of which was raised through donations from companies such as Boeing, McDonald's and Wrigley, and from private individuals. The park's green spaces are complemented by works by renowned artists and architects. Particularly noteworthy are the huge steel concert shell with seats for around 11,000 visitors by star architect Frank Gehry, the "*Cloud Gate*" artwork and Crown Fountain. Cloud Gate is a 20 m high and 11.5 million US dollar stainless steel sculpture in the shape of a cloud in which the Chicago skyline and the people standing nearby are reflected (**Fig. 5.13**). The park's biggest attraction is the Crown Fountain, created by Spanish artist Jaume Plensa and donated by

ity with a children's museum, a greenhouse, a 15-story Ferris wheel, an IMAX theater, restaurants, retail, and a ballroom, and a wraparound promenade, and became an instant visitor magnet (Spirou 2006, p. 297, 2011, p. 282).

The southern part of the Central Area was particularly unattractive with many brownfields and dilapidated buildings, but had great development potential due to Grant Park. In the 1980s, plans were initiated to convert some of the brownfields on the Near South Side, which had long been used by Central Station, which closed in 1972, and by warehouses, into residential areas. In 1993, the first neighborhood, named Central Station, opened with luxury townhouses in what was then a desolate setting. To make a statement and draw attention to the new downtown residential location, Mayor Daley himself moved into the new neighborhood (Bennett 2010, p. 78). At the turn of the millennium, the City of Chicago redeveloped the aging Grant Park and renovated or in some cases expanded several museums located at its south end, transforming them into a museum campus. The adjacent Soldier Field football stadium to the south and the McCormick Place fair and exhibition grounds have

Fig. 5.13 "Cloud Gate" in Millennium Park

5

◘ **Fig. 5.14** "Crown Fountain" in Millennium Park

◘ **Table 5.1**   Visitor numbers of selected attractions in Chicago (in million)

Attractions	1995	1999	2010
Navy Pier	3.0	7.7	8.7
Millennium Park[a]	–	–	4.5
John G. Shedd Aquarium	1.84	1.85	2.1
Art Institute of Chicago	2.2	1.4	1.8
Willis Tower observation deck[b]	1.4	1.3[c]	1.36
The Field Museum	1.26	1.5	1.2
Viewing platform Hancock Tower	k. A.	k. A.	0.6

Sources: Clark et al. (2002, p. 504); Chicago Office of Tourism and Culture (2011)
[a]Opened in 2004
[b]By 2009 Sears Tower
[c]1998 figures

the Crown family, which is formed by two 15 m high towers facing each other at a distance of about 77 m (◘ Fig. 5.14). The towers are clad in glass blocks that act as LED-controlled video screens. Water flows constantly from the roof of the towers during the summer and, in addition, powerful jets of water come out of the mouths of the faces of Chicago citizens projected onto the towers at intervals of a few minutes. The water flows into two shallow pools between the two towers, with which Plensa wanted to give the impression that it was possible to walk on water. On hot summer days and nights, the water offers cooling, especially for children, who run screaming back and forth between the water-spouting towers (Plensa 2008).

In Millennium Park, attractions financed by donations have been created that can be used free of charge by wealthy and poor visitors alike. This also applies to the numerous concerts on the open-air stage of the Gehry Pavilion. Many works of art invite visitors to join in and visibly enjoy themselves. At the same time, globally marketable icons have been created. Although Millennium Park visibly enhances downtown Chicago, it is often criticized because it is not a public space, but a private and highly controlled and regulated one. Moreover, it is feared that the fun society will soon demand something new and even more exciting, and wear and tear will be inevitable even with the best of care. The polished stainless steel cloud is already not as shiny as it was in the first few years after completion. Navy Pier had lost its appeal after the opening of Millennium Park and is undergoing refurbishment and partial remodelling only 20 years after its opening (author's observations; Calbet i Elias et al. 2012, p. 392; Jayne 2006, p. 192; Spirou 2011, pp. 282–285). Navy Pier and Millennium Park are not the only recreational facilities that have sprung up in downtown Chicago in the past two decades (◘ Table 5.1). In the *Loop*, older stages have been renovated or reopened

for theaters or shows, and north of the Chicago River, numerous restaurants draw visitors late into the evening. The Chicago Art Institute, one of the finest museums in the U.S., added a modern art wing in 2009. Since the 1990s, a new downtown has emerged in which leisure and entertainment offerings play a major role (Newman and Thornley 2005, p. 105).

5.3.3 Reurbanisation

Downtown Chicago has all the hallmarks of reurbanization. For decades, the city has focused on the redevelopment of the center and has been adept at mobilizing private capital for revitalization. The Central Area has developed far more positively than probably even the boldest optimists had hoped in the early 1990s (◘ Fig. 5.15). From 2000 to 2010 alone, the number of inhabitants increased by 36% to 182,000 and that of the *Loop* by as much as 187% to 20,280. In the *Loop*, more than 8000 new residential units were created not only in high-rise buildings but also in older office buildings that were converted. The development south of the *Loop* is particularly noteworthy. Until the end of the nineteenth century, the wealthiest citizens of the city had lived here (Zorbaugh 1929). After they moved to the Gold Coast, the magnificent mansions fell into disrepair. The houses of the former super-rich on Prairie Avenue have been renovated in recent years and sold for prices in the double-digit millions. Numerous high-rises with hundreds of new apartments were built in the neighborhood. The number of jobs has been more or

new millennium, tourist numbers had risen by around 10%, only to collapse back to their old levels as a result of the recession. In 2010, the occupancy rate was low at only 69.6% given the many new hotels (Chicago Office of Tourism and Culture 2011, p. 1). In 2010, only 1.13 million international tourists visited Chicago. After the much smaller Boston, Chicago ranked only tenth on the list of most popular cities with foreigners (▶ www.census.gov). The city has not yet succeeded in attracting the attention of international tourists. The planned Olympic Games in 2016 could have contributed to global marketing, but unfortunately Chicago did not win the bid.

Today, downtown Chicago is considered the region's flagship. Mayor Richard M. Daley has responded to the demands of a globalized world and set the right accents in city policy (Simpson and Kelly 2011, p. 218). Nevertheless, one can also criticize the development. Apartments have been built almost exclusively for the rich and the super-rich, the number of families with children is low, and there has been a privatisation of public space. In the 1970s, however, the inner city was conceivably unattractive and perceived as dangerous. Today, it is clean and crime is low; downtown's quality of life is higher than ever. What is particularly encouraging is that, starting from the *Loop*, reurbanization is now spreading out in a circle like annual rings around a tree and has even encompassed the long-neglected neighborhoods south of the *Loop* and west of the Chicago River. What is worrying, however, is that almost all efforts have been focused on revitalizing the inner city, while the problems of the rest of the city have been neglected (Newman and Thornley 2005, p. 106).

In Chicago, soft location factors are now more important than industrial expansion. Policymakers have created publicly accessible goods such as clean air, safe public spaces and parks, and attractive recreational offerings. This is where people like to stay and spend their money (Clark et al. 2002, pp. 510–512). The entertainment industry has become the city's growth engine, and at the turn of the millennium the leisure and cultural sector was the city's most important industry. Consumerism is clearly at the center of it all. Once America's leading industrial city, it has become a consumer city. Even in the face of declining population in the city as a whole, Chicago is once again brimming with confidence. In 2003, the city adopted a Central Area Plan that defines very ambitious goals for the next 20 years: *"The Central Plan is a guide for the continued economic growth, and environmental sustainability of Chicago's downtown for the next 20 years. It is driven by a vision of Chicago as a global city, the 'downtown of the Midwest', 'the heart of Chicagoland', and the 'greenest' city in the county"* (City of Chicago 2003, Introduction). Given the successes since the early 1990s, at least the new urbanites

☐ **Fig. 5.15** Trump Tower, hotel and apartments next to Wrigley Building

less stagnant this decade due to the recession starting in 2008, but was impressively large at just under 500,000 in the Central Area and about 300,000 of those in the *Loop* in 2010. The number of hotel rooms in the *Loop* has increased from 8038 to 10,639 in 10 years. The *Loop* is home to 16 institutions of higher education, including DePaul University and Roosevelt University, with a total of 65,000 students. The retail sector has also seen positive growth. While State St. in the *Loop* is home to more trendy stores for young people such as Old Navy and Urban Outfitters, North Michigan Avenue is home to international designers (Chicago Loop Alliance 2011). It cannot be ruled out that the number of residents and jobs in the Central Area will continue to increase. Although Chicago gives the impression of great density to European visitors, there is still enough brownfield or underutilized land available to be developed by 2020. Moreover, some of the buildings that no longer meet today's standards will be replaced by (mostly taller) new buildings (City of Chicago 2003, p. 31).

However, the development of tourist numbers is not satisfactory. Although the city has strongly promoted and expanded the cultural sector and many new hotels have been built, just under 40 million foreign guests visited Chicago in 2000 and 2010. At the beginning of the

living downtown do not doubt that these goals will be realized. The impoverished citizens on the South Side and the West Side are likely to be less optimistic about the future.

5.4   Boston and the "Big Dig" Project

Boston is one of the oldest and most traditional cities in the USA. In 1630, 11 ships with around 700 Puritans from England docked in Massachusetts Bay in search of a place where they could develop freely. Boston soon developed into a center of shipbuilding, fishing, and whaling. The Puritans' pursuit of education continues to have a positive effect today. In 1636 a college was founded in neighboring Cambridge and in 1639 was named Harvard after its founder. By 1730, the city had a population of 13,000. Tensions between the colonists and the mother country led to the famous Boston Tea Party in 1773. After the USA gained independence in 1776, Boston expanded its port and trade and was officially elevated to a city in 1822 with a population of just over 40,000. From the mid-nineteenth century onwards, Boston developed into a location for the food and leather industries, shipbuilding, mechanical engineering and the steel industry in the course of industrialisation. The education sector was not neglected either. In 1848 the first public library in the country and in 1861 the Massachusetts Institute of Technology were opened in Boston. At the same time, the population grew rapidly with the immigration of poor Irish who began arriving in large numbers in 1846. Chinese and Eastern European Jews also immigrated in the late nineteenth century. By 1900, Boston had a population of nearly 560,000. Like other East Coast cities, Boston suffered from deindustrialization beginning in the mid-twentieth century. The population peaked at 801,000 in 1950, only to decline to 563,000 by 1980 due to suburbanization. At the same time, some neighborhoods began to decline sharply (Blume 1988, pp. 369–375; Kennedy 1992, pp. 43–71).

In the past decades, a structural change has succeeded. Boston was able to develop into the leading service center of New England and had 617,600 inhabitants again in 2010 (▶ www.census.gov). At the same time, a process of upgrading downtown and adjacent areas began. Since the 1970s, this has included the revitalization of Quincy Market and transformation into a popular *festival market* (◘ Fig. 5.16), the construction of the Prudential Center with several hotels and a large shopping mall, and the *gentrification of* the residential neighborhoods near downtown Beacon Hill and Back Bay (◘ Fig. 5.17). More recently, the downtown area has become much more attractive with the relocation of the Central Artery. This courageous (and expensive)

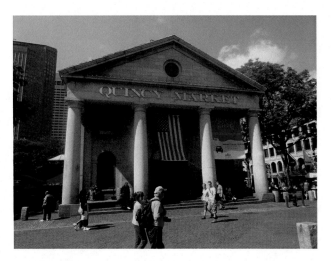

◘ **Fig. 5.16**   Quincy Market

◘ **Fig. 5.17**   View over the Back Bay (Boston)

construction measure shows how important the commitment and assertiveness of planners and politicians are for the realization of spectacular large-scale projects.

5.4.1   Central Artery

The Massachusetts shoreline is glacially influenced and is characterized by a close interlocking of land and water. In the nineteenth century, land fill and shoreline changes occurred with the expansion of the harbor, and continued through the twentieth century (Miller 2002). At no time was downtown Boston, located on a peninsula, well connected to the rest of the city. The problems increased with the rise of personal transportation, and traffic became increasingly congested on the narrow streets of downtown. As urban development in the mid-twentieth century was guided by the model of the car-oriented city, Boston hoped to improve traffic flow and at the same time increase the attractiveness of the city by expanding its streets. In 1948, an ambitious Master Highway Plan was adopted for the Boston metropolitan area, which included the construction of the Route 128 beltway and the Central Artery, an elevated urban highway through the city center. The intent was to improve access to downtown and free the streets from moving traffic. Although early protests arose against the plans, construction began in 1950. More than 100 residential buildings and 900 businesses fell victim to the wrecking ball. Thirty-four on and off ramps were to ensure maximum accessibility, but took up more space than the Central Artery itself. The North End, which was completely cut off from the city center, China Town and the Leather District were particularly affected by the construction and demolition measures. At the time, the (partial) demolition of problematic neighbourhoods in need of redevelopment was not uncommon as part of infrastructure measures, as it was hoped to kill two birds with one stone. The Central Artery, which was opened to traffic in 1959, soon turned out to be a complete mistake. The many ramps did indeed cause many motorists to use the new elevated road, but it quickly reached its capacity. Instead of contributing to an upgrading, the elevated road had led to a further loss of attractiveness of the city center (Aloisi 2004, pp. 6–11).

5.4.2 Realisation

A paradigm shift began when Democrat Michael Dukakis became governor of Massachusetts in 1975 and he appointed Fred Salvucci as secretary of transportation. Both felt that further expansion of the inner-city road network would do more harm than good to the city and that Boston's quality of life needed to take center stage. At the time, the Central Artery, planned for 75,000 vehicles per day, was used by more than twice that many. Salvucci's idea was to put the downtown elevated highway underground and build a third tunnel under the harbor to relieve traffic congestion. Not another house was to be destroyed. Old neighborhoods were to be united and downtown green spaces created, and Salvucci hoped that the Federal Highway Administration (FHWA) would cover up to 90% of the construction costs. However, the project initially became a pawn in partisan politics. Governor Dukakis was succeeded in 1979 by Edward J. King, who was also a Democrat, but considered the development of transportation routes to the airport, which is located on an offshore peninsula, a top priority. All access roads to the airport ran through East Boston, where many homes and a recreational area were to be demolished or sacrificed. In light of the failed Central Artery, protests grew in Boston against the construction of more roads and the planned destruction of East Boston. Opponents received support from the charismatic Democrat Tip O'Neill (1912–1994), whose constituency was East Boston. O'Neill was a member of the House of Representatives in Washington for 34 years and its speaker since 1977. Due to opposition to the project and other unfortunate circumstances such as the imprisonment of the Massachusetts Secretary of Transportation for bribery while King was in office, O'Neill's plans were not implemented. Dukakis was again governor of Massachusetts from 1983 to 1991. In Washington, however, the takeover of funding for the Boston transportation project now failed because of Republican President Reagan (1981–1989). Democratic Senator Edward Kennedy of Boston finally managed to get a number of Republicans to side with the Democrats and vote for the project in 1987. Now that the FHWA had secured the assumption of a large part of the costs, concrete planning could finally begin (Aloisi 2004, pp. 16–27).

An important goal of the Central Artery/Tunnel project, popularly known as the "*Big Dig*," was to improve the quality of life. Neighborhoods dissected by the elevated highway were to be reunited, public transportation was to be improved, and 120 ha of open space and green space were to be created. On the site of the former elevated highway, a new inner-city park, the Rose Kennedy Greenway, was to be created on the roof of the tunnel. The greenways promised attractive access to the waterfront, which previously could only be reached after crossing under the elevated highway (Aloisi 2004, pp. 22–32). Due to a lack of alternatives, traffic had to continue to be routed through downtown Boston throughout the construction period. The tunnel was built first, and only after its completion could the old elevated highway be demolished. Construction of the multi-level tunnel was difficult because the Financial District high-rises to the right and left had to be shored up before excavation could begin. The underground expressway was opened to traffic in late 2003 (◻ Fig. 5.18). At times, up to 5000 workers were involved in the construction, which consumed up to $3 million per day. A year later, dismantling of the old elevated highway and construction of the planned Rose Kennedy Greenway began. A major problem was the planned tunneling of the Fort Point Channel on the east bank of the Charles River, which was estimated to cost $1.5 billion. Fort Point Channel is an unattractive marshy area on whose banks Gillette produces more than a billion razor blades a year. Tunneling the production facilities would have jeopardized the precision manufacturing of the blades. To get around these problems, they built a bridge over the canal. In 2002, the ten-lane Bunker Hill Memorial Bridge, designed by architect

5

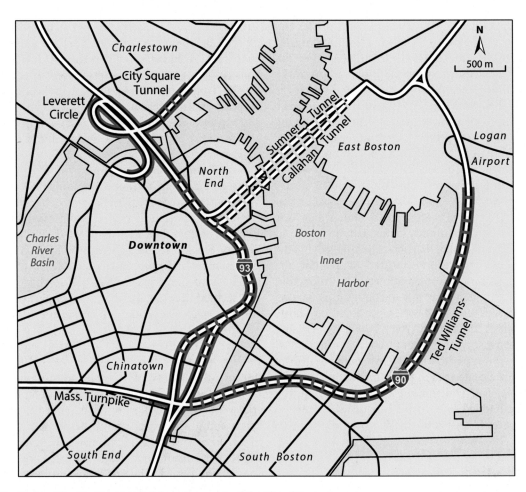

■ **Fig. 5.18**   Boston: *The Big Dig*. (Adapted from documents of the City of Boston)

■ **Fig. 5.19**   The Zakim Bridge is considered the new landmark of the city of Boston

Leonard P. Zakim and known in Boston as the Zakim Bridge, opened (■ Fig. 5.19). The suspension bridge connects the new downtown tunnel to Interstate 93 and U.S. Highway 1 and is considered a new landmark in the city (Aloisi 2004, pp. 43–68). The "*Big Dig*" project,

including construction of a third tunnel to the airport, demolition of the old elevated highway, and creation of the Rose Kennedy Greenway, was completed in late 2007 after 15 years of construction. The total length of all new lanes in tunnels, on the bridge and access ramps is 270 km. The cost of the project was $24.3 billion, far higher than initially calculated (Miller 2002; Associated Press 10/7/2012a).

5.4.3 Evaluation

As is often the case with major projects, there were many problems during construction and after completion, such as falling ceiling parts, non-functioning lighting in the tunnels, and a surface on the access ramps that was not suitable for New England's winter temperatures, which were discussed at length in the press. Still, there is no doubt that downtown Boston and especially Spectacle Island, which fronts the city and had previously served as a dumping ground, have become enormously more attractive. Most of the excavated material for the tunnel construction had been dumped on Spectacle Island.

Fig. 5.20 The elevated Central Artery used to run in the foreground of the image. (After demolition of the elevated road, the North End awoke to new life)

The island was transformed into a recreational area with 28,000 newly planted trees and shrubs, biking and hiking trails, beaches, a pier with moorings for 38 boats, and a dock for the ferry from Boston (Aloisi 2004, p. 32). In downtown Boston, the Central Artery had presented an almost unbridgeable bar between downtown and the waterfront; today, the Rose Kennedy Greenway provides recreational space and opens the city to the water. The North End has once again become part of downtown Boston and is barely recognizable just a few years after its completion (◻ Fig. 5.20). Not long ago, the houses showed severe signs of decay and many vacancies. In the meantime, sidewalk cafés attract tourists and signs of *gentrification* with all its advantages and disadvantages are visible (author's observation).

5.5 Detroit: A Doomed City?

The rise and decline of the city of Detroit can be divided into three phases. From 1910 to the end of the 1940s, the city grew very rapidly; the following three decades were characterized by strong suburbanization; and since the end of the 1970s, Detroit has suffered from competition from other regions of the United States (Hill and Feagin 1987). Among shrinking cities, Detroit occupies a special place. With about 1.3 million people, the city has lost more residents since 1950 than any other city in the United States, and there is no end in sight to the shrinking process. At the same time, Detroit is characterized by a high poverty rate and unemployment, as well as a low level of education among the population. In more recent times, there have been a few bright spots. However, any hopes of averting the city's ultimate demise were dimmed when Detroit filed for bankruptcy in June 2013.

Detroit was founded in 1701 as a French fort on the river of the same name, which connects Lake Erie with the western Great Lakes, by the French Captain Cadillac, and was controlled by the French until 1760. The opposite side of the Detroit River remained in French hands and has been part of the newly formed Canada since 1867. In the mid-nineteenth century, the city had only about 20,000 inhabitants, but experienced a first boom in the course of industrialization from 1850, when ovens and other kitchen appliances were built here. Detroit also became known as a location for breweries and confectionery. In 1900, the city already had 285,000 inhabitants and ranked 13th among American cities (▶ www.census.gov).

5.5.1 City of the Automobile

The decisive development push was triggered by Henry Ford, who built his first automobile in Detroit in 1896. Other pioneers of vehicle construction were the Dodge brothers, Packard, Studebaker and Chrysler. The city also saw the emergence of an extensive subcontracting industry for automobile manufacturing. By 1920, Detroit's population had risen to nearly one million, climbing to fourth place behind New York, Chicago, and Philadelphia (▶ www.census.gov). The period of unbridled growth went hand in hand with the introduction of assembly line work and mass production by Henry Ford. In 1913, the modern industrial age dawned with the opening of the factory in the enclave of Highland Park, some 10 km from the center of Detroit (◻ Fig. 5.21). Ford had succeeded in building an automobile for everyone with the "*Model T*" and changed America. By 1927, when production of the Model T ceased, 15 million vehicles rolled off the assembly line.

Fig. 5.21 Ford factory in Highland Park. (The assembly line was invented here in 1913)

5

The population of Highland Park increased from only 427 inhabitants in 1900 to 52,000 in 1930. Since production was carried out on several levels at the Highland Park plant, but Ford wanted to combine all production steps at one location and there was no room for expansion, he opened the River Rouge plant in Dearborn, south of Detroit, in 1918, where he had an 800 m long assembly line built. The new plant even had its own blast furnaces and was the largest industrial complex in the world at the time. Inland freighters could deliver ore and coal from northern Michigan. The Highland Park plant closed in the late 1950s (Lacey 1987, pp. 9, 114).

Assembly line work brought Detroit several decades of prosperity and growth, but it also ushered in its decline. Uneducated workers could be quickly trained and had little need for knowledge. Because each worker was responsible for only a small portion of production, they could not accumulate greater knowledge. Detrimental to the region's innovative power was the fact that of the initially many producers, only the *big three* Ford, General Motors and GM remained. Detroit developed into a monostructured city, with hundreds of thousands of poorly educated workers employed by just three major automakers, which in the long run stifled competition and worker creativity. To make matters worse, the city never invested much in expanding its education sector (Glaeser 2011a, pp. 48–58). Beginning in the mid-twentieth century, problems began to accumulate in Detroit. In the decade following the end of World War II, the *big three* built 20 new automobile plants exclusively in suburban areas, where supplier industries and populations also shifted. Wages in the auto industry had always been high, but now the powerful United Auto Workers (UAW) union was imposing unreasonably high wages that hurt Detroit. Wages and living standards of industrial workers in Detroit were at times higher than in any other American city. But blacks were reluctant to be hired, were allowed to work only the dirtiest jobs, and were disadvantaged in the housing market (Sugrue 1996, pp. 41–47, 91–110). It is not surprising that several days of race riots broke out in 1967, some of the most brutal in the United States, and encouraged the exodus of whites from the city. Those employed in the auto industry were predominantly white and lived in well-kept single-family homes in the suburban area, while the impoverished black population remained in the core city and the building stock visibly deteriorated. Although workers in the auto industry earned comparatively well, bolstered by the UAW, they repeatedly went on weeks-long strikes, forcing GM to shut down production for 113 days in 1945–1946 and 54 days in the late 1990s. In addition, auto manufacturers underinvested. By the early 1970s, Detroit's manufacturing plants and infrastructure were outdated, taxes and wages high, and workers egged on by unions aggressive. Under these conditions, it was

understandable that capital moved to the southern U.S. or abroad, and no new investment occurred in Detroit (Gallagher 2010, p. 37; Hill and Feagin 1987; Ross and Mitchell 2004, pp. 687–688).

5.5.2 Population and Building Stock

Detroit's population fell 57% from 1970 to 2010, to just under 714,000, while that of the suburban area rose 27% to 4.3 million. In the core city, blacks made up only 16% of the population in 1950; in 2010, they made up 82.7%. Detroit and the enclave of Highland Park suffer from the fact that, unlike most cities, immigrants from Mexico, China, or other countries are not moving in. In both cities, residents are poorly educated and poverty is high. In 2009, at the height of the last recession, nearly a third of the population was unemployed. The figure is probably still understated, since many people stopped looking for work long ago and are no longer listed in the statistics. More than half of Detroit's children live in poverty. Half of adults can barely read and only 2% of all public school students can read well enough to qualify for college. Detroit has one of the highest crime rates in the U.S. and the murder rate is eight times higher than New York, but the percentage of murderers caught is far lower in Detroit. The situation is quite different in neighboring Dearborn, the site of the River Rouge factory (◻ Table 5.2). Many people from Arab coun-

◻ **Table 5.2** Socioeconomic data for Detroit, Highland Park and Dearborn 2010

	Detroit	**High- land Park**	**Dear- born**	**USA**
Inhabitants	713,777	11,664	97,144	308 million
Blacks (%)	82.7	93.5	4.0	13.1
Born abroad	5.1	0.4	25.5	12.8
Minimum Bachelor's degree Persons over 25 (%)	12.2	9.0	30.3	28.2
Average per capita income in US dollars	15,216	11,756	22,816	25,482
Poverty rate (%)	36.2	47.5	25.0	15.7
Unemployment rate 2011 (%)	19.9	25.9	8.0	8.9

Source: ▶ www.census.gov

tries and Europe have moved to Dearborn in recent decades, and about one-third of residents now have Arab roots (Benfield 2011; Deskins 1996, pp. 259–260, 280; Gallagher 2004, 2010, pp. 119–120; Malanga 2010, p. 5; Sugrue 2004, p. 230).

There are hardly any blocks of cohesive development left in Detroit, and signs of decay are visible everywhere (◘ Fig. 5.22). The low value of the remaining residential buildings has a negative impact on the city's tax revenues, as property taxes fell by nearly 20% from 2008 to 2013 alone. In 2013, there were 78,000 vacant buildings in Detroit, about half of which were classified as dilapidated or uninhabitable (◘ Fig. 5.23). In addition, there were 66,000 properties for which no owner felt responsible anymore (The Economist 22/6/2013). The establishment of a *land bank* was long controversial. It was not until federal funds were made available for a *Neighborhood Stabilization Program* in 2008 that Detroit established a *land bank*, but it did not

really work. Only a scant 10% of the brownfield land that becomes city property is put to new use by investors. Buyers have to take care of the land registration themselves after acquiring the land in an often difficult process. In addition, the city of Detroit often makes mistakes in the sale of brownfield sites. It is even said to have happened that the land was sold to an investor, but the ownership rights went to someone else (Gallagher 2010, pp. 36, 141). Countless brownfields are now dominated by the grey-green of prairie grass, and it seems as if nature is reclaiming the city. The brownfields are larger than in other cities because not only has the population left the city, but so have many of the automobile and supplier companies that once occupied huge tracts of land. Many of the lots have lain fallow for decades and are completely overgrown, and some of the residential buildings have already collapsed. At the beginning of the twentieth century, the town had grown rapidly in only a short time. The quickly constructed wooden houses were of poor quality and weathered quickly. The high water table contributed to moisture seeping into the houses. Detroit is referred to as *urban prairie* or the term new suburbanism is used to describe the city (Gallagher 2010, pp. 21–30; Hanlon 2010, p. 55). At 350 km², Detroit's parish is large compared to that of other cities, and its population density is low at 2037 people per km². All of Boston, San Francisco, and Manhattan, with a combined population of about three million, would fit within Detroit's parcel (◘ Fig. 5.24). However, most suburbs have much lower population densities than

◘ **Fig. 5.22** Dilapidated residential buildings in Detroit

◘ **Fig. 5.23** Not very trustworthy apartment building in Detroit

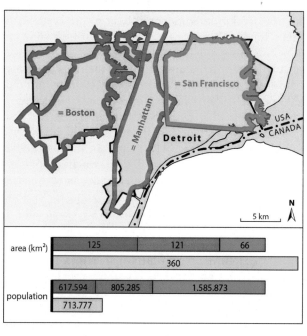

◘ **Fig. 5.24** Detroit parish compared to other cities. (Adapted from Gallagher 2010, p. 30)

Detroit. Moreover, the suburbs were planned as only loosely populated areas and today have far higher tax revenues than Detroit (Gallagher 2010, p. 12). ˘

5.5.3 Revitalisation Efforts

The first efforts to revitalize downtown were made in the 1970s. Henry Ford II, the grandson of the company's founder, believed that a huge office building with the programmatic name Renaissance Center on the banks of the Detroit River could save the city. He succeeded in attracting a number of other investors and architect John Portman to build the huge complex. Consisting of seven buildings, the Renaissance Center was completed in 1981 at a cost of $350 million. At 221 m, it is still the tallest building in the city, though it has been much criticized for its fortress-like character, and for putting a bar between downtown and the river. Because there was a lot of empty office space downtown in the 1980s, the Renaissance Center could not be fully leased. In 1996, it was sold for a mere $100 million to General Motors, which had the interior completely remodeled and put its world headquarters and a Marriott hotel there. A glassed-in multi-story atrium opening onto the Detroit River is very attractive today, especially since a beautiful promenade has been built along the river in the meantime. GM's sales exhibit also attracts visitors, while the storefronts, which were laid out on several levels, are partially empty. After hours and on weekends, the Renaissance Center's public spaces are deserted (Gay 2001, p. 11; Glaeser 2011a, p. 62; Lacey 1987, pp. 373–395; Plunz 1996). Another bad investment was the nearly 5 km long monorail built in the 1980s with public funds at a cost of $200 million, which runs on stilts in a ring around the city center and connects the most important buildings. Although the ride costs only 0.5 dollars, the People Mover is only used by around 6500 people a day. The construction was unnecessary anyway, as the wide streets below are always empty. The monorail has to be subsidized with 8.5 million US dollars annually (Glaeser 2011a, p. 62). In the late 1990s, the city of Detroit approved the construction of three casinos in the city center, which have since opened with associated hotels and parking garages. Visitors to the gambling venues rarely leave, however, and a look around the casinos gives the impression that this is where the city's poor gamble away what little money they have. However, the downtown area does come alive for a few hours on weekends or in the evenings when the Detroit Lions play at Ford Field football stadium, the Detroit Tigers at Comerica Park baseball stadium or the Detroit Red Wings at Joe Louis Arena. Then even the suburbanites venture into downtown Detroit in large

■ **Fig. 5.25**   Compuserve building in downtown Detroit

numbers. Many downtown restaurants are open only on game days (Meyer and Muschwitz 2008, pp. 33–34; Ross and Mitchell 2004; author's observations).

On a more positive note, some downtown office buildings have been built in more recent times. In the early 1970s, three young people started Compuserve in the suburban area. In the late 1990s, the city of Detroit succeeded in attracting the company to Detroit with tax breaks and a promise to use only Compuserve software. In 2002, Compuserve opened its 15-story headquarters in the center of downtown with just over 100,000 sq. ft. of office space (▶ www.compuserve.com) (■ Fig. 5.25). Further jobs were created in the city center when the 13-storey office building One Kennedy Center opened in the immediate vicinity in 2006 with tenant Ernst & Young. The two buildings are located at the intersection of Woodward and Michigan Avenues, where there had formerly been a park, but later the unsightly central bus station. After the bus station was moved to another location, a public-private partnership between the City of Detroit and several businesses reopened Martius Park in 2004 with very nice green space and an ice skating rink in the winter. Martius Park and the new River Walk waterfront on the Detroit River are the most attractive places in downtown Detroit, which continues to be characterized by many brownfields and vacant buildings. Detroit-born businessman Dan Gilbert caused a stir in 2010 when he moved the jobs of 1700 employees of his company, Quicken Loans, from the suburban area to downtown. Quicken Loans is the largest online mortgage provider in the U.S.. A year later, 2000 more employees moved. Meanwhile, Gilbert has invested more than $300 million in the redevelopment of nine downtown office buildings. Shops are to be installed on the ground floor of the buildings he has acquired. Gilbert wants a vibrant downtown that combines the functions of living and working. He knows that well-educated people are essential to a recovery. The conversion of the Broderick Tower, built in 1928 on Woodward Avenue, into an apartment building by investor Motown Construction Partners also offers hope. The 35-story building, which opened in 1928, had been vacant for decades. In late 2012, the first tenants moved into the 125 apartments in the renovated building. Even the penthouse was easy to rent out for a monthly rent of $5000. Other former office buildings are currently being converted into apartment buildings in Detroit. Investors are rewarded with generous tax breaks (Forbes 9/18/2012; Gay 2001, pp. 37–39; Huffington Post 4/5/2012; The Economist 10/22/2011).

While there are some bright spots downtown, the economy of the city as a whole is performing more than sluggishly. From 1970 to 2000, the city lost about 165,000 of 230,000 industrial jobs, but gained only 30,000 in services. In the 1980s, Detroit subsidized the construction of Chrysler Motors' Poletown plant on the city's east side, but only a few thousand jobs were created there. Some 1500 homes and 144 businesses had been demolished to build the new plant. Some 3500 mostly elderly people had lived in the once white enclave (Malanga 2010, p. 4; Sugrue 2004, pp. 230–234). On a more positive note, Wayne State University took an initiative based on the edge of downtown. In 2003, General Motors donated a vacant building to the university for a business incubator to provide space for young companies. The facility has been well received and generously supported by private donors. By fall 2009, 111 small companies had already set up shop in TechTown. The spectrum of newly founded companies is extraordinarily broad, ranging from electrical engineering and biotechnology to a company that restores old paintings from South America. Unlike other business incubators, the leases are not time-limited. Tenants are even encouraged to stay as long as possible (Gallagher 2010, p. 123).

After decades of mismanagement and political scandal, the city found hope in 2009 when Dave Bing became mayor. Bing had been one of the best basketball players in the U.S. in his youth and then successfully managed an initially ailing steel mill for three decades. Bing sought to rebuild Detroit along the lines of a private company. He stipulated that City Hall employees must be neatly dressed and on duty by eight in the morning as a visible sign that the slovenliness was over. A panel of experts appointed by Bing found ways to save about $500 million from the city budget by, for example, privatizing airport management and downsizing the administration. Detroit's schools are among the worst and most expensive in the US. Although the population has been declining for decades, the number of teachers has not been reduced, although an estimated 500 are not doing their jobs. In order to identify these unknowns, each employee must now collect their wages in person from the agency. In addition, Bing is seeking to raise $200,000 from private funds to establish at least 70 *charter schools* in the city (Malanga 2010, p. 7).

Mayor Bing has recognized that Detroit cannot be saved by expensive construction, but open space must be used wisely. Bing wants to concentrate the remaining population in selected locations to create vibrant neighborhoods again. About a quarter of Detroit's land area is to be converted into parks and into forests (Glaeser 2011a, p. 66; Malanga 2010, pp. 6–7). Implementing this plan is difficult. First, it must be decided which neighborhoods should be razed to the ground and which should be revitalized. In any case, demolishing houses and leveling entire neighborhoods is expensive. At the same time, new homes must be built elsewhere. Since the city has little tax revenue, it has to rely on federal funds, which is difficult. Problematically, many residents want to stay in their homes, even if only one or two houses on their block are still occupied. Evictions and forced relocations are legally difficult or even impossible (Associated Press 8/3/2010). Plans to shrink Detroit back to health are also met with criticism. Binelli (2011) argues that Detroit should not shrink, but grow by incorporating neighboring communities. If Detroit were to incorporate these suburbs, many problems would be solved. The proposals are unrealistic, however, because the wealthier communities are unlikely to be interested in merging with the problem-plagued city. The state of Michigan lacks the legal basis for forced incorporations.

5.5.4 Outlook

Downtown Detroit has developed unexpectedly well in recent years thanks to the commitment of private investors. But vacant buildings and derelict sites still far outweigh in the city center. Things look even worse for the city as a whole. Mayor Bing has good ideas, but their implementation is hardly financially feasible. With tax revenues low, Detroit relies on private donors. The nonprofit organization Greening of Detroit has planted about 3000 trees every year for more than 25 years and supports urban gardens. The Detroit RiverWalk was funded with donations from the Kresge Foundation, which has also supported the revitalization plan for the city as a whole (Gallagher 2010, pp. 144–145; The Economist 10/22/2011). All of the successes to date have unfortunately been a drop in the bucket, and the private chapter is limited. One must doubt that the city's extensive problems can be solved. So far, the population decline has not been halted. In March 2013, the state of Michigan hired Kevyn Orr, an attorney specializing in bankruptcy law, to clarify Detroit's financial circumstances. Orr determined that the city had debts of $17 billion, or $25,000 per resident, and was insolvent. The bankruptcy is hitting the city and residents very hard, as salaries and pensions can no longer be paid in full, and the already ailing infrastructure and poorly equipped police and fire departments, as well as schools, are facing further cuts. It is to be expected that even more people will turn their backs on Detroit (The Economist 22/6/2013).

5.6 Miami: Economic Recovery of a Polarised City

Miami, previously the site of a fort to protect against Spanish attacks, was founded in 1896 with a population of only 300 and was connected to the railroad in the same year. Since the west of South Florida is occupied by the swampy Everglades and the eastern part does not have a good natural harbor or resources, only a few people settled there at first. The region developed into a destination for wealthy tourists from the north who enjoyed spending the winter in the warmth of Florida. Particularly popular was an island just a few hundred meters wide and about 30 km long with wide sandy beaches off Miami. The southern part of the island is occupied by Miami Beach, where many hotels in the Art Deco style were built in the 1920s and 1930s. But it was not until the 1950s, when air conditioning became common in private homes, that more and more people moved permanently to the region (Grosfoguel 1995, p. 158 f.; Nijman 2007, p. 102).

With the takeover of power of the communists by revolutionary leader Fidel Castro in 1959 on the island of Cuba, which is located only 180 km from the southern tip of Florida, the boom of South Florida began. To date, more than three million Cubans have left their homeland in several waves. From 1959 to 1962, about 200,000 Cubans, mostly upper class and upper middle class, emigrated to the United States. These Cubans were well-educated, often academics or entrepreneurs, and were least able to come to terms with the new regime. In the U.S., they received a positive reception and generous financial support. Almost all Cubans settled in south Florida, and here preferably in Miami. The climate of South Florida was similar to that of Cuba, and the proximity to the homeland was seen as an advantage. The refugees hoped that the U.S. would take effective action against Fidel Castro and that their stay in the U.S. would be short. After the failure of the U.S. invasion of the Bay of Pigs in 1961, however, these hopes evaporated. To prevent an over-concentration of Cubans in southern Florida, the U.S. funded resettlement to other parts of the country in the early 1960s. When the finan-

cial support ended, however, most Cubans returned to Florida, where they settled preferentially west of the city (Boswell 2000, pp. 155–157; Rothe and Pumariega 2008, p. 250). In the mid-1960s, private property was nationalized in Cuba and opponents of the regime were subjected to repression. As Fidel Castro sought to rid himself of critics, he allowed Cubans who had relatives in the United States to leave the country. However, young men under 27 and well-educated Cubans of certain professions were largely exempt. *Freedom flights* did not cease until 1973, after some 250,000–300,000 Cubans had fled to the United States (Boswell 2000, p. 145; Perez 1986; Rothe and Pumariega 2008, p. 251). The third wave of emigration began in 1980 as demonstrations against the communist government increased. Fidel Castro labeled the insurgents *"scum"* and allowed their departure on U.S. boats. 124,000 Cubans, mostly working class, who had once cheered communism, left the island disappointed. This included, for the first time, a larger number of black Cubans. Castro seized the opportunity to declare some 2000 inmates of prisons and mental institutions, as well as prostitutes, to be opponents of the revolution and let them leave as well. Miami-Dade County immediately saw a sharp rise in crime. In addition, conflicts arose between Cubans and blacks, who had experienced a modest economic boom in the preceding years as a result of the civil rights movement. Working-class Cuban immigrants became competitors in the labor market with blacks, who were also mostly poorly educated. While the immigrants of earlier waves of immigration had been welcomed with open arms in the United States, the Cubans arriving now were not very welcome. The fourth wave of immigration was triggered by the collapse of the friendly Soviet Union, when communist Cuba suffered an economic setback and civil war nearly ensued. In 1994, Castro allowed 37,000 Cubans to leave the island on boats and rafts. As more and more of the Cubans remaining on the island fled to the US base at Guantanamo Bay, Castro and US President Bill Clinton agreed to allow 20,000 Cubans a year to enter the U.S.. These were no longer considered political refugees but economic refugees and received no financial support (Boswell 2000, p. 147; Rothe and Pumariega 2008, pp. 252–253).

5.6.1 Population Structure

Miami-Dade, the southernmost county in mainland Florida, is identical to Greater Miami. The population increased fivefold from 495,000 in 1950 to 2.5 million in 2010. A good 70% of the growth is based on migration gains. The proportion of Hispanics has risen from just 4% in 1950 to 64.5% in 2010, half of whom are Cuban.

The proportion of Cubans is declining, however, as more and more people migrate from other Central and South American countries, most of whom are less educated than Cubans. In 1970, just over 90% of all Hispanics living in Greater Miami were still Cubans. Colombians and Nicaraguans are now the second and third largest groups, accounting for 7.5% and 7%, respectively. Non-Hispanic whites and non-Hispanic blacks now account for only 16.0% and 19.3% of Greater Miami's population, respectively (Croucher 2002, pp. 225–226; Portes and Stepick 1993, pp. 150–175; ▶ www.census.gov).

Greater Miami today is highly segregated ethnically and socioeconomically (◘ Fig. 5.26). Miami, the largest city with a population of 400,000, has suffered from *white flight*, the migration of middle and upper class whites from the cities to the suburbs, for decades (Croucher 2002, p. 228). In the city, Hispanics have not had to adjust to the English-speaking population because of their large numbers. Miami only tem-

porarily became a bilingual city; for decades Spanish has been the most common language. Non-Hispanic whites have moved to suburban areas or neighboring northern counties. This is now true for many successful Cubans as well. At the same time, increasing numbers of Hispanics from other Central and Latin American states have moved into the city of Miami, forming their own neighborhoods such as Little Haiti or Little Managua. In 2010, 70% of Miami residents were Hispanics, but only 28% were Cubans (▶ www.census.gov). Cubans' preferred place to live is Hialeah, the second largest city in the region with 218,000 residents. In Hialeah, Cubans make up 72% of the population. Non-Hispanic blacks are concentrated in northwest Miami-Dade County, and non-Hispanic whites prefer the offshore islands and the community of Coral Gables. Ethnic segregation corresponds with socioeconomic segregation. In 2010, the median per capita income in Greater Miami was $23,348. In Brownsville, where 74.7% of the popu-

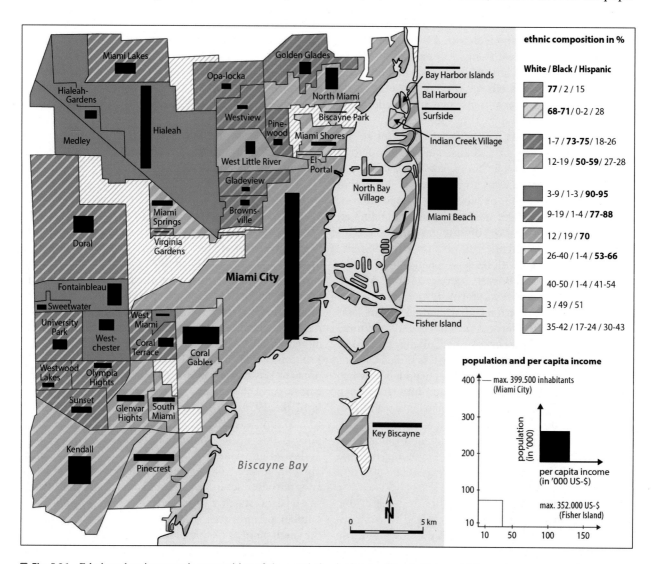

◘ **Fig. 5.26** Ethnic and socioeconomic composition of the population in Greater Miami. (Data and Maps ▶ www.census.gov)

lation is non-Hispanic black, it was only $12,880. This ethnic group had the least chance of upward mobility because they could not benefit from the international-ization of the economy (see below). In Key Biscayne, with a 36.5% non-Hispanic white population, per capita income was $70,000. Overall, Greater Miami suffers from a highly polarized population and lacks a middle class. Educational attainment is poor and household income is among the lowest in the United States. The proportion of non-Hispanic whites has been declin-ing in recent years; but it is whites who have been the region's sustaining middle class. Poor Hispanics and blacks relatively rarely succeed in moving up into the middle class. (Brookings Institution 2004; Nijman 2000, p. 140; ▶ www.census.gov).

The southern tip of Florida has attracted not only political refugees and economic migrants from the Caribbean and Latin America, but also a cosmopoli-tan elite that prefers to maintain second homes near the coast and on the offshore islands. Often these people live in *gated communities* or own well-guarded apartments in the high-rises of Miami Beach. Fisher Island and Indian Creek Village, with only 132 and 86 residents, respectively, are particularly well-guarded and are both only accessible by private boats. The almost exclusively non-Hispanic residents, who are said to include many international actors and wealthy athletes, have a far above-average per capita income (▶ www.census.gov).

Without the influx of Cubans, the Miami region's rapid growth and economic boom would not have been possible because there were no endogenous locational factors to attract capital or labor (Portes and Stepick 1993, pp. 204–205). The well-educated and bilingual Cubans of the first two waves of immigration found opti-mal conditions for rapid integration in the United States. They were welcomed with open arms and given gener-ous financial support (Boswell 2000, p. 248). In Miami, the Little Havana district was formed, which functioned as a catchment area for the Cuban immigrants of later waves (◘ Fig. 5.27). Networks emerged that facilitated the entry of all later arrivals. Many Cubans chose to be self-employed and benefited from the less skilled Cubans of the later waves of immigration as cheap labor, which allowed them to be quickly integrated into the Cuban community. Cubans maintained their identity and pro-vided the best possible support for newcomers. In 1985, Cubans assumed political leadership in Miami when Xavier Suarez was elected Cuban mayor for the first time (Mohl 1983, p. 73; Nijman 2007, p. 100; Rothe and Pumariega 2008, p. 254).

The large number of communities and the strong ethnic and socioeconomic segregation lead to competi-tion between the individual groups. A melting pot whose residents pursue similar interests has not emerged in

◘ **Fig. 5.27** Mural in Little Havana

Greater Miami (Gainsborough 2008, p. 420). Just over half of the region's residents are foreign-born and have little identification with where they live. Its location at the southern tip of Florida, inter-group competition, and lack of roots are cited as reasons for a crime rate that has been very high for decades. At the time of Prohibition, rum from Havana was smuggled into the USA via Miami. In the 1970s, the city became an impor-tant gateway for drug smuggling from Latin America. Today, the legal and illegal sectors are strongly inter-linked and cannot always be clearly assigned (Census Bureau 2011b; Nijman 2007, p. 101).

5.6.2 Post-industrial *Global City*

Compared to most other U.S. cities, Miami has been late in gaining greater prominence. New York, founded in the early seventeenth century, was already established in the pre-industrial phase. Chicago was elevated to city status in 1833 and is considered a prototype industrial city. Los Angeles, which is often used as an example of a post-industrial city, experienced its greatest growth between 1900 and 1920. Nijman (2000, pp. 137–139) therefore refers to Los Angeles as an ex-industrial city and Miami, whose most important growth phase did not begin until the 1950s, as the prototype of a post-industrial city. Miami was never an industrial city, but developed directly into an internationally oriented service center. Today, Miami-Dade County has an important mediat-ing role between North and South America and exer-cises partial functions of a global city. In the Caribbean, small island states predominate, in which only a limited number of industries and services can develop due to the small number of inhabitants. Individual states there-fore concentrate on certain functions. A hierarchical city system based on the division of labor has developed in the Caribbean, with Miami at the top of the hierarchy. The city performs important control functions and is the hub for people, goods and money of the Caribbean region (Grosfoguel 1995).

Already in the late 1920s, the first American airline Pan Am had its hub in Miami and offered flights to Havana and other destinations in Latin America. Miami expanded this function after the Second World War for passenger and cargo traffic. Miami International Airport offers passengers direct flights to 64 countries in Central and South America and the Caribbean. That's more than from any other airport in the United States. Even New York and Los Angeles offer fewer direct flights to foreign countries than Miami. Miami Airport is also the most important U.S. hub for cargo traffic to and from Latin America and the Caribbean, with 83% of imports and 80% of exports handled here coming from or going to these regions. Computers and other technical equipment predominate among exports, while flowers top the list of imports, followed by fish and vegetables (Miami International Airport 2010). The Port of Miami has comparatively little importance due to its peripheral location on the North American continent and the lack of water transportation options to other regions of the U.S.. The U.S. ports in the Gulf of Mexico have far better conditions. With a throughput of only 6.7 million tonnes of freight, the Port of Miami ranked only 60th among US ports in 2009, but in terms of container throughput Miami ranked 14th (► www.aapa-ports.org).

Many Latin American countries are politically unstable and national banks are considered unsafe. It is therefore not surprising that wealthy residents of Latin America and the Caribbean prefer to deposit their money in Miami banks and invest in real estate in the fast-growing region. Miami has even been voted the "*best place to do business in* (!) *Latin America*" several times (Nijman 2007, p. 101). In 2011, 14 foreign banks of mostly Latin American origin maintained a branch in Greater Miami. This is far fewer than in New York and Los Angeles, but more than in Chicago with only six branches. The banks profit from drug trafficking and other illicit businesses, as a not insignificant portion of profits are said to be deposited here or invested in real estate in South Florida (Nijman 2007, p. 101). An important feature of *global cities* is the control of the world economy exercised through headquarters of large corporations. Miami has a deficit in this area. In 2012, only two of the 500 largest U.S. companies, World Fuel Services and Ryder System, were headquartered in Greater Miami. They also ranked only in the back of the pack, at 85th and 407th respectively (► www.fortunemagazine.com). Nevertheless, there is no doubt

that Miami is now the most important business location in the Caribbean. Conducive to this is the fact that in Florida union influence and taxes are low and therefore labour is comparatively cheap (Girault 2003, p. 29; Nijman 1997, 2007, p. 96; Sinclair 2003, p. 219).

Miami has strong cultural ties to the Spanish-speaking countries of the Americas and has been called the "*Hollywood of Latin America*". While Los Angeles is the undisputed center of the feature film industry, Miami produces television shows, especially telenovelas. In addition, several major networks such as TV Azteca, which serves the entire Spanish-speaking world, are based in Miami. In addition, Miami benefits from immigrants from different regions who have brought with them the typical music of their homeland. In the meantime, Miami has developed its own style, which is broadcast to many countries via television stations such as MTV Latino. Miami's creative potential is enormous. This is reflected in the establishment of many dotcom companies, mostly in attractive beachfront locations. The location is referred to as Silicon Playa in reference to California's Silicon Valley (Sinclair 2003). Miami is also one of the most important Internet hubs. The Terremark Building in downtown Miami, which the Internet service provider Verizon purchased for $1.4 billion in early 2011, is the hub of more than 160 global networks, making it reportedly one of the five largest Internet hubs in the world (Blum 2011).

5.6.3 Tourism

Tourism is an important industry in Miami. In 2011, 13.4 million tourists visited the city and the adjacent coastal resorts, of which 12.5% were business travellers. Just under half of all guests came from abroad. Visitors from South and Central America and the Caribbean represent the largest group of international visitors, followed by travelers from Europe and Canada. Tourists spend an average of 5.82 days in the region and spend an astonishing $264 per day. A good 40% stay overnight in Miami Beach, which has a high proportion of luxury-class hotels (◘ Fig. 5.28) (Smith Travel Research 2012). The Art Deco style hotels at the southern tip of Miami Beach are a particular attraction (◘ Fig. 5.28). In the 1970s, the old buildings were in very poor condition and the neighborhood was almost demolished. Fortunately, this was prevented and most of the buildings have since been rehabilitated. In addition, many new hotels have been built in a similar style, so it is often difficult to distinguish between old and new buildings. Wealthy tourists and the high percentage of cosmopolitan residents contribute to Miami Beach's high purchasing power and high-end retail.

Fig. 5.28 Art deco building in Miami Beach

5.6.4 Future Prospects

Miami is benefiting greatly from the growing interest in cruises. In 2009, the port of Miami counted 4.1 million passengers (1999: 3.1 million), far more than any other port in the USA. Most of the cruises are to destinations in the Caribbean, but some also go to South America. A positive factor for Miami as a business location is that some of the world's largest cruise companies, such as Carnival Cruise Line, Royal Caribbean International and even Norwegian Cruise Line, have their headquarters in Miami. The majority of cruises take place during the winter months, as the hot and humid summer climate and frequent hurricanes are uninviting during the summer and fall (Port of Miami 2010a, b). Although the settlement of Cubans was not always seen as an advantage, and the U.S. even engaged in (less than successful) relocation efforts at times, it was the emergence of an enclave Hispanic economy that enabled Miami's rise as a mediator between North and South America. Greater Miami's advantage was that it could guarantee great political and economic stability compared to most of Latin America and the Caribbean. There was much to be said for depositing or investing money in Miami. Miami hopes to further develop its role as a mediator between North and Latin America and is actively promoting the creation of a free trade area to encompass the entire Americas. In 2003, the city hosted the summit to prepare the *Free Trade Area of the Americas* (FTAA). Should this be realized, Miami would like to become the site of the organization's secretariat (Gainsborough 2008, p. 422).

5.7 Atlanta: The Suburban Space as a Location for Services

The first whites settled at the later site of the city of Atlanta in 1823. The boom began in 1837 when the Western and Atlantic Railroad, coming from Tennessee, extended its route. The small settlement at the terminus of the railroad was first called Terminus but was soon renamed Atlanta, which gained political independence in 1847. The city, which lacked natural locational factors such as mineral resources or a good position on a waterway, was able to develop into an important commercial and logistical center in the following decades with the support of far-sighted decision-makers and inexpensive labor. Early warehouses were established in Atlanta for cotton and foodstuffs from the South and manufactured goods from the North. By 1890, 11 railroads served the city, and Atlanta increasingly became a hub of commerce between northern and southern states and a major site of wholesale and retail trade. Atlanta came to the attention of national investors when it hosted the International Cotton Exposition in 1891. In the years that followed, a number of textile mills opened in Atlanta; however, the city never developed into a major industrial location (Ross et al. 2009, pp. 51–52; Sjoquist 2009, p. 4).

5.7.1 Upswing to a Service Center

In the twentieth century, Atlanta consistently expanded its role as a transportation hub and goods distribution center. In 1925, the construction of the airport laid an important foundation for the further development of the region. In the 1990s, the number of passengers at Hartsfield International Airport surpassed that of Chicago O'Hare, the country's largest airport to date, for the first time. Today, Atlanta's airport has by far the largest passenger volume in the United States. With the expansion of the U.S. interstate highway system since the 1950s, three interstate highways (I-85 from the East Coast to Alabama, I-20 from the East Coast to Texas, I-75 from Michigan to the southern tip of Florida) intersect in Atlanta. What was once a rail hub has become a hub for goods and people by rail, road and air. Around 80% of the U.S. population lives no more than 2 h by plane or a day on the highway from Atlanta. Atlanta was able to expand its importance as a logistics center when a U.S. Customs Inland Port was built in the city as part of the preparations for the 1996 Olympic Games. Since then, containers from abroad can be transported duty-free to Atlanta. With deepwater ports opening promptly on Georgia's Atlantic coast in Savannah and Brunswick, Atlanta's container throughput surged in the mid-1990s. The Summer Olympics also enhanced Atlanta's image and contributed to its global marketing (Hartshorn 2009, p. 135; Ross et al. 2009, pp. 52–74).

The Atlanta region has also experienced an upswing in recent decades due to the changed political climate in the southern states. Since the U.S. South was characterized by racial problems until the 1960s, potential investors avoided Atlanta for a long time. This locational disadvantage no longer exists today (Ross et al. 2009, p. 72). The comparatively new and good infrastructure as well as low labor costs and land prices are major advantages of the region. Atlanta has been able to develop from a city of regional importance to a second-tier *global city* in the past decades. Local politicians like to compare Atlanta with New York or Chicago; in fact, however, Atlanta is in many ways more like Boston, Denver or San Diego. In 2010, 13 of the 500 largest U.S. companies were headquartered in the Atlanta region, 11 of which were in the core city. Several of these companies, such as UPS (ranked 52nd) and Delta Airlines (ranked 53rd), are involved in logistics and transportation. Home improvement chain Home Depot ranks 35th; but Coca-Cola (ranked 59th) may be the world's best-known company by far (▶ www.fortunemagazine.com). In addition, the CNN television network has its headquarters on the edge of downtown Atlanta. Even if the channel is not one of the world's 500 largest companies by revenue according to Fortune magazine, it enhances the city's global image as news is broadcast around the world from here 24/7. The number of people employed in upper management is growing at a similar rate as the workforce in basic services. In addition, many people are still employed in the transport sector, in warehouses and in wholesale trade. Fewer people are working in manufacturing, but the share of workers in retail, hotels, and financial and health services has increased. At first glance, Atlanta is not a stronghold of national and international tourism. Nevertheless, its function as a hub of goods and people and a large number of trade shows and conventions have always attracted many visitors to this convenient location, and the 1996 Olympics led to an increase in the number of hotel rooms to about 60,000 in the region (Jaret et al. 2009, p. 45; Newman 1999, pp. 268–288; Ross et al. 2009, p. 57).

Today, Atlanta is the capital of the state of Georgia and the economic center of the southern states. Hardly any other region in the USA has grown as rapidly in recent decades as Atlanta MA, whose population rose from 1.7 million in 1970 to 5.2 million in 2010. Population growth has been confined exclusively to suburban areas, as the core city's population has declined

from nearly 496,000 to 420,000 in 2010 over the same period. In 1990, the lowest level had been reached with only 394.000 residents. As the population of core Atlanta has rebounded since then, optimists believe a further increase to 530,000 residents in 2030 is possible. Overall, however, the percentage of people living in the core city has declined significantly. This trend is expected to continue over the next few decades, as the outer counties are currently growing particularly rapidly. While about 29% of MA residents lived in the City of Atlanta in 1970, only 8.1% did so in 2010. As suburbanization has increased, population density in the region has been cut in half from 1970 to the present, to just 570 per square mile, and is not as low in any of the other 15 largest metropolitan areas in the U.S. as it is in Atlanta MA. Lots for single-family homes average 0.3 acres, and each in-migrant now occupies twice as much land as it did in 1970, more than 0.2 acres. Since 1950, Atlanta MA has expanded from three to 20 counties. The core area is the *10-county region* (Cherokee, Clayton, Cobb, DeKalb, Douglas, Fayette, Fulton, Gwinnett, Henry, Rockdale) with a population of about 4.1 million in 2010 (Atlanta Regional Commission 2011; Hartshorn 2009, p. 150; Ross et al. 2009, pp. 53–55; ▶ www.census.gov).

5.7.2 Decentralisation

Although the population in the suburban area had already been growing much faster than in the core city for some time, politicians and planners still assumed in the 1980s that new office jobs would be created almost exclusively in the center of Atlanta. They thought they were well prepared, since all streets ran in a star shape toward the center and it was easily accessible from all directions. The euphoria increased even more when the Metropolitan Atlanta Transit Authority's (MARTA) plans to build several rapid transit lines took shape (Hartshorn 2009, p. 142). In fact, however, jobs increasingly migrated to suburban areas. Retail took a lead in this process, but soon new office space was also built almost exclusively in suburban areas.

By the mid-twentieth century, Peachtree St., located in downtown Atlanta, was the region's main shopping street, although many retailers had set up shop on arterial streets to serve customers who drove their cars to suburban areas. The days of people shopping on foot in close proximity to where they lived were soon gone for good. While goods used to come to customers, the private vehicle made it possible for customers to drive to the goods. The cost of transportation has passed from the retailer to the customer; at the same time, companies have been able to cut costs due to economies of scale such as reduced storage capacity. In return, the customer accepts long distances and, therefore, time and gas costs. In Atlanta, retail initially preferred to relocate along the connecting road between downtown and Buckhead, which is located to the north (Hankins 2009, pp. 87–91). Buckhead, which today forms a triangle between the intersection of I-75 and I-85, had been established in the nineteenth century north of the city as a residential neighborhood for the wealthy and was incorporated by Atlanta in 1952. To this day, large lots with imposing homes in the plantation style typical of the southern states characterize the district, because the upper class never left the park-like Buckhead with its curved streets and dense tree population. In any case, extremely lush vegetation is characteristic of Atlanta, which is probably the greenest city in the USA (■ Fig. 5.29). The high purchasing power in Buckhead and the good inter-regional transport connections were probably the most important reasons why Lenox Square Mall, one of the first enclosed shopping centers in the USA, opened at this location in 1959 with a lettable area of 62,000 m² (■ Fig. 5.30). Lenox Square was a great success and the location was so good that 10 years later a second shopping center of similar size was built in the immediate vicinity on Peachtree Road, Phipps Plaza. Soon, the first shopping centers were built in Atlanta's suburban area, favoring convenient locations and the high-income communities south of the city of Atlanta. The southern part of the region is less attractive as a residential location because it is home to Hartsfield International Airport, one of the largest airports in the world, and also many warehouses and industrial buildings. In 1964, the Columbia Mall opened in DeKalb County 11 km east of Atlanta with 32,000 m² of leasable space. The shopping center was never a great success, was rebuilt several times and renamed Avondale Mall after a total

■ **Fig. 5.29** View from Stone Mountain over Atlanta, probably the greenest city in the USA

renovation. However, as customers failed to come, the center was demolished in 2007 (International Council of Shopping Centers 2012)).

Until the beginning of the 1970s, shopping centers were primarily built along the ring road that runs around the city about 10 km from downtown Atlanta. Since the beginning of the 1980s, more and more areas beyond the beltway have been developed for housing estates. As retail follows the population in the suburban area or, in some cases, large-scale operators buy and build on land very early, shopping centers have sprung up at an ever-increasing distance from the core city. Most shopping centers that have opened since the 1980s wrap around Atlanta like a second ring a good 20 miles away. The Mall of Georgia, which remains the largest shopping center in the region with a leasable area of almost 160,000 m², was even built around 55 km from the center of Atlanta (■ Fig. 5.31). When the Mall of Georgia opened in 1999, it was located at the far northern edge of the urban region. However, it was foreseeable that suburbanization would expand beyond the Mall of Georgia in the coming years, and indeed it soon did. Shopping centers are under a great deal of competitive pressure and must constantly adapt to the spirit of the times. The bigger and fancier the offerings, the more attention shoppers pay to shopping centers. The Lenox Mall has been renovated several times in the past decades and most recently expanded to 140,000 m². The individual shopping centers try to satisfy different target groups and needs. Some of the centers cater to utility shopping, while others emphasize experiential shopping. The former are rather plainly built and try to attract as many customers as possible by offering a low price. The North Point Market Center, which just opened in Fulton County in 2012, is a good example of a so-called *power center*. Discounters like Babies "R" Us, Marshalls, and PetSmart offer low-priced merchandise here in large freestanding retail spaces called *big boxes*, with direct access from the parking lots. The much more elaborate Mall of Georgia, with 18 movie theaters and several themed restaurants, tries to appeal not only to shoppers but also to leisure-oriented customers. Phipps Plaza in Buckhead appeals to a particularly affluent audience with branches of the luxury department store chains Saks Fifth Avenue and Nordstrom as magnets and retailers such as Hugo Boss, Tiffany and Valentino (Hahn 2002, pp. 57–141; ICSC: Shopping Center Database).

The suburban area of the Atlanta region has not only become increasingly important as a residential and retail location, but also for offices. By the mid-1970s, there were already as many office jobs in the suburban area as in the core city of Atlanta (Hartshorn 2009, p. 144). Especially along the I-285 beltway, large office buildings

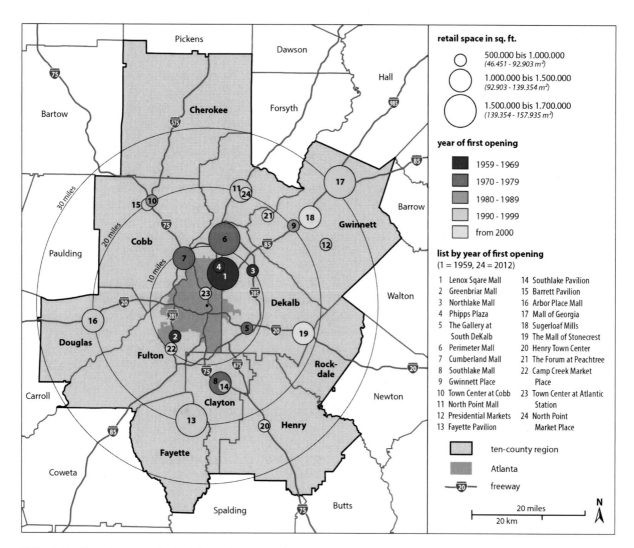

Fig. 5.30 Shopping centers in the Atlanta 10 county region. (Data and map basis ▶ www.census.gov)

The map legend contains:

retail space in sq. ft.
- 500.000 bis 1.000.000 (46.451 - 92.903 m²)
- 1.000.000 bis 1.500.000 (92.903 - 139.354 m²)
- 1.500.000 bis 1.700.000 (139.354 - 157.935 m²)

year of first opening
- 1959 - 1969
- 1970 - 1979
- 1980 - 1989
- 1990 - 1999
- from 2000

list by year of first opening (1 = 1959, 24 = 2012)

1 Lenox Sqare Mall
2 Greenbriar Mall
3 Northlake Mall
4 Phipps Plaza
5 The Gallery at South DeKalb
6 Perimeter Mall
7 Cumberland Mall
8 Southlake Mall
9 Gwinnett Place
10 Town Center at Cobb
11 North Point Mall
12 Presidential Markets
13 Fayette Pavilion
14 Southlake Pavilion
15 Barrett Pavilion
16 Arbor Place Mall
17 Mall of Georgia
18 Sugerloaf Mills
19 The Mall of Stonecrest
20 Henry Town Center
21 The Forum at Peachtree
22 Camp Creek Market Place
23 Town Center at Atlantic Station
24 North Point Market Place

ten-county region
Atlanta
freeway

Fig. 5.31 The Mall of Georgia is one of the largest shopping centers in the USA

and office parcs were often built in close proximity to shopping centers, and *edge cities* sprang up. Data on the number of *edge cities* in the Atlanta region vary. Garreau (1991, pp. 426–427) identified four fully developed *edge cities* and three other locations that were still in the process of becoming *edge cities*. Lang (2003, p. 134) has demonstrated only two *edge cities*, but Hartshorn (2009, pp. 147–148) has demonstrated four *edge cities*, the latter emphasizing that only Buckhead has a residential population of any significant size (Table 5.3). The differences in the designation of *edge cities* arise due to definitional imprecision and because precisely defined boundaries are lacking. The concept of *edge cities* has become less important in recent decades as offices have increasingly been built in isolated locations. About 60% of the region's workforce works in *edgeless cities*, but only 40% in *edge cities* (Hartshorn 2009, p. 151).

5

□ **Table 5.3** Employees in *edge cities* in the Atlanta MA (in thousands)			
Edge city	**1990**	**2000**	**2005**
Perimeter city	183	231	196
Cumberland	83	122	109
Buckhead	75	100	96
Roswell/Alpharetta	34	108	138
Total	375	561	539

Source: Hartshorn (2009, p. 147)

5.7.3 City Center

Parallel to the construction of shopping centers, retail in downtown Atlanta has suffered major losses. Whereas in Europe retail performs an important guiding function in downtowns and attracts thousands of customers every day, downtown Atlanta is now almost exclusively home to office buildings, hotels and vacant storefronts. Since Rich's and Macy's, which was in need of renovation anyway, closed in 1991 and 2003, respectively, there are no department stores left downtown. Smaller stores are also hard to find (Hankins 2009, p. 93), despite early efforts to make downtown more attractive. In the early twentieth century, an underground business and entertainment district had been created in conjunction with the construction of a railroad bridge, but had closed by the late 1920s. In 1969, Underground Atlanta reopened with small shops, restaurants, and nightclubs and was initially very successful, benefiting from the fact that alcohol could be served here far more freely than at other locations in the region. Underground Atlanta lost its appeal when the serving of alcohol was relaxed throughout the region. In 1980, Underground Atlanta closed, but reopened in 1989 after a major renovation. Never truly successful, Underground Atlanta has faced closure several times in more recent times. Semi-popular, Underground Atlanta is only popular with tourists who stray downtown because of the nearby The World of Coca-Cola museum or Centennial Olympic Park (Rice 1983, p. 41). More successful was the opening of Atlantic Station on 56 acres on the edge of Midtown Atlanta in 2005. Atlantic Steel had formerly produced steel at this site. The old manufacturing buildings were converted for mixed use with retail, recreational facilities, and loft-style apartments. Homes were also built on the former plant site. The mix of different uses is intended to allow for day and nighttime activity. Not only the old buildings remind of the former use, but also several large

steel sculptures, which should create a unique identity and serve the better marketing. Architects, urban planners and investors celebrated Atlantic Station as a development reminiscent of former urban life. The amenities of suburban living, i.e., homes as townhomes or duplexes with manicured lawns and private security, are ostensibly combined here with city amenities such as pedestrian-friendly shopping and numerous stores including a supermarket and movie theaters. Atlantic Station's retail, however, is not different from that of suburban shopping centers, with the same chains offering identical products. In addition, since 2007, efforts have been underway to revitalize the traditional main shopping street of Peachtree St. over a 14-block length. This section of the street is being converted into a pedestrian zone, and it is optimistically hoped that a good 90,000 m^2 of new retail space will be created here. Since the 1996 Olympic Games, the number of inhabitants in the city center has increased considerably, but in view of the great competition from the many shopping centers in the region, it is doubtful whether the old center of Atlanta will be able to develop into a supra-regional retail location again. Today, downtown primarily performs the function of an office, hotel, and convention location (Hankins 2009; Hartshorn 2009, pp. 146–149).

5.8 New Orleans After Hurricane Katrina

New Orleans markets itself as the *Big Easy* and stands for jazz, voodoo romance, casinos and a pleasant life. The colourful Mardi Gras attracts hundreds of thousands every year. But appearances are deceptive. The city suffers from social problems, there is a lack of jobs in upscale services or in forward-looking industries, and due to its location in the delta of the Mississippi River, New Orleans is one of the most vulnerable places in the U.S.

Unlike most other cities in the south of the country, New Orleans looks back on a comparatively long history and has a world-renowned historic core with the French Quarter. The city was founded in 1718 by the French in the delta of the Mississippi and named after the Duke of Orleans. It passed to Spain in 1762 and to the United States in 1803 as part of the *Louisiana Purchase*. However, not only the French and Spanish were involved in the early settlement of the region, but also British traders and many African slaves who served as laborers in the early stages. Later, the city grew in several waves through German, Irish, Italian and other groups of immigrants (Fussell 2007, p. 847). Because of its location at the mouth of the Mississippi River, New Orleans grew into a major port city early on. The agricultural products of the Midwest were brought across the Mississippi River to the port of New Orleans, where many products for the central United States arrived at the same time. By 1840, New Orleans was the third largest city in the U.S. after New York and Baltimore, which, however, had just over 100 more people. In the twentieth century, with the expansion of the canal, road, rail, and pipeline networks and the containerization of freight, the port lost importance as a locational factor. In 1960, New Orleans reached an all-time high population of 627,000, but ranked only 15th among U.S. cities. Whites made up more than half the population at the time. Due to suburbanization, which disproportionately involved the mostly comparatively wealthy whites, the number of residents dropped to only 484,000 by 2000, about two-thirds of whom were black. These often belonged to lower socio-economic groups, as many of the better-educated blacks had left New Orleans for Atlanta or Los Angeles. The average household income at the turn of the millennium was about a quarter below the American average. Accordingly, the poverty rate and the number of welfare recipients in New Orleans were above average. As in many other cities, ethnic groups were highly segregated. A few very exclusive neighborhoods with an affluent and predominantly white population contrasted with a large number of poor neighborhoods with very poor building stock. The Garden District, located on the high banks of the Mississippi River, represents an island of affluence with its large colonial-style houses. Far more typical of New Orleans is the Desire district, made famous by Tennessee Williams' 1947 drama *A Streetcar Named Desire*. The proportion of blacks here was 94.1% in 2000. The population in Desire was younger and the number of female-headed families was far greater than in the Garden District. These values correlated with high poverty and low household income in Desire. The poor neighborhoods were located in the deeper and thus more flooded parts of the city. Aside from the petroleum processing plants in southern Louisiana, there are no significant industries in the region. Many people work in low-paying services such as those provided by transportation, hotels, restaurants, and other tourist facilities. There are also a disproportionate number of jobs in public administration (Gelinas 2010; Hahn 2005b; Hirsch 1983; Vigdor 2008, pp. 138–141; ▶ www.census.gov).

5.8.1 Hurricane Katrina

The Mississippi River had repeatedly overflowed its banks throughout history, creating a natural dam about 4 m high, on which the French laid out their settlement with a checkerboard layout (Ford 2010, p. 13). Adjacent to the north of this causeway was a marshy area that sloped down towards Lake Pontchartrain, some 2.5 km away. The marshes provided an ideal breeding ground for pathogens and needed to be protected from further flooding before settlement could occur. The raising of dykes and the construction of drainage canals had begun by the French in the eighteenth century, but it was not until the nineteenth century that canal construction progressed according to plan. With growing settlement pressure, the entire terrain between the Mississippi River and Lake Pontchartrain was drained and settled after World War II. Drainage of the marshes caused further subsidence of the land that was once regularly flooded (◘ Fig. 5.32). In addition, the town was flooded several times during violent storms when the levees broke. For years, scientists had pointed out that the poorly maintained levees would not withstand a hurricane and that a major disaster would one day occur. Ironically, a book by Craig Colten (2005), describing in great detail the history of levee construction and settlement development in New Orleans, with all its dangers and negative effects, was published just as Hurricane Katrina was destroying the city.

In the last days of August 2005, the coast of the states of Louisiana, Mississippi and Alabama was hit by a hurricane with extreme wind speeds. As the storm changed direction before making landfall and lost power slightly, people in New Orleans initially breathed a sigh of relief. But when several levees to Lake Pontchartrain broke, the lake's waters flooded about 80% of the city, and New Orleans, which was largely below sea level, filled up like a giant bathtub (◘ Fig. 5.33). Before the hurricane hit New Orleans, there had been warnings about the monster storm, and eventually orders to evacuate homes had been issued. Not all of the city's residents had heeded the call, however. Some were afraid of looting, others didn't want to leave their pets behind, or didn't have a car or gasoline with which to leave town. As the water rose higher and higher, some people climbed onto the roofs of their homes or made last-minute trips to the comparatively high-altitude Superdome sports stadium, where conditions were disastrous. Since 1979, the Federal Emergency Management Agency (FEMA) has been responsible for disaster relief, which was placed under the newly formed Department of Homeland Security (DHS) after the terrorist attacks of 2001. During the New Orleans flood disaster, FEMA failed completely. The agency was underfunded and relied on the cooperation of a large number of private companies and non-profit organizations. Because action had to be taken very quickly, the wrong partners were often chosen, and many contracts were completed sloppily or at inflated costs (Gotham 2012, pp. 636–637). Rescue efforts got off to a slow start, but eventually almost the entire population was evacuated. The poorer people were bussed to Houston,

5

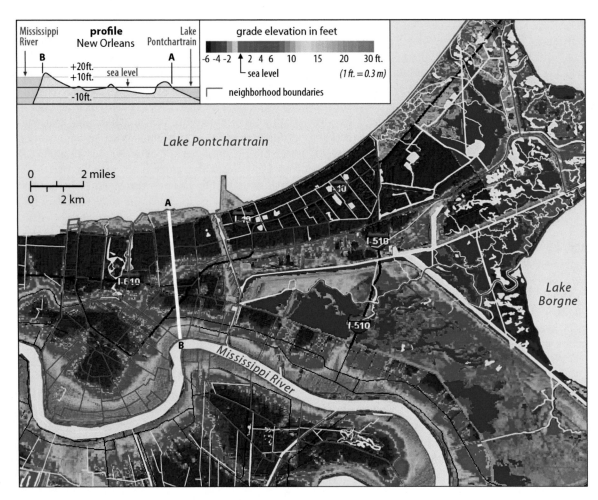

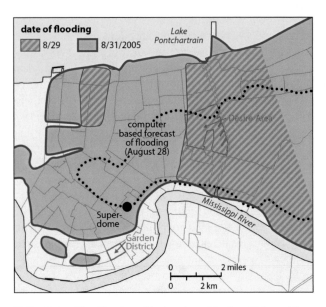

□ **Fig. 5.33**   Flooding caused by hurricane Katrina. (Adapted from New York Times 3/9/2005; Hahn 2005b, Fig. 1)

while many of the slightly wealthier stayed with relatives, friends, or in hotels around the country. Hurricane Katrina not only largely destroyed New Orleans, but also neighboring parishes. Neighboring St. Bernard was even more damaged than New Orleans. Sadly, nearly 1600 people in southern Louisiana lost their lives to Hurricane Katrina. Many homes were under water up to the roof ridge, and it took weeks to pump out the dirty water. Mismanagement of relief efforts and looting exacerbated the situation. It soon became clear that some of the lower lying parts of the city were completely destroyed. The flooding had destroyed two-thirds of the approximately 215,000 housing units or rendered them uninhabitable due to extensive damage (□ Fig. 5.34). Neighborhoods near the Mississippi River, such as the French Quarter and the adjacent Downtown with many office and hotel buildings, as well as the Garden District, were hardly affected because they were located on or near the high bank that forms the edge of the water-filled basin (□ Fig. 5.35). Here, at most, the roofs had

▫ Fig. 5.34 Destroyed residential buildings in New Orleans

▫ Fig. 5.35 Undamaged house in the Garden District (March 2006)

been removed or the windows had been crushed by the wind. Above all, the houses in lower locations, where the socially weak often lived in rented houses, had been destroyed. Whole neighbourhoods were no longer habitable, and housing had become a scarce commodity. The poorer population in particular could not return because they had lost everything. Insurance companies were reluctant to pay out, as many people were insured against storm damage but not against water damage. Renters were especially poorly served, as only home-owners were compensated. Even some of the wealthy, whose homes were barely damaged, did not return because they did not trust the promised reinforcement of the levees or did not want to pay the now very high insurance sums for their homes (Hahn 2005b; Vigdor 2008, pp. 146–147). In New Orleans, it was clearly demonstrated that the new forms of organization created by neoliberalism fail when disaster strikes. Since private

companies can only operate successfully if they make a profit, victims become clients who are helped not on the basis of social criteria but after weighing up costs and benefits (Gotham 2012, p. 635).

5.8.2 Reconstruction

The 2010 census counted 343,829 inhabitants in New Orleans, roughly the same number as in the 1910 census. On the one hand, this was gratifying, because even this comparatively low figure had not been expected for a long time, since the residents who had fled or been forcibly evacuated before the hurricane had initially been very slow to return. On the other hand, nearly 29% fewer people lived in the city in 2010 than 10 years earlier, and New Orleans had fallen to 52nd place among U.S. cities. The number of residents in New Orleans MA fell from 1.36 to 1.235 million, or by 11%, from 2000 to 2010. Of the 100 largest *metropolitan areas in* the U.S., only 11 experienced a population loss during this time period; however, no other MA lost nearly as many residents on a percentage basis as New Orleans (Greater New Orleans Community Data Center 2011, p. 3). Regardless of the level of destruction, nearly all neighborhoods in New Orleans have experienced population losses. Hurricane Katrina also changed the socioeconomic and ethnic makeup of the population. Since far more blacks than whites had lost their homes, they were less likely to return to the city. After decades of decline, the proportion of whites has risen again (▫ Table 5.4).

There was, of course, no plan in New Orleans in 2005 for rebuilding the city in the event of near-total destruction, and those responsible were overwhelmed when the water finally drained away and the full extent of the damage became apparent. For several months, there was serious debate as to whether New Orleans should be rebuilt or completely abandoned (McDonald 2007). The levees were in very poor condition and very expensive to rebuild. It was doubted whether they would ever be able to protect the city in the event of another major hurricane. The population had been declining for decades, schools and jobs were poor, the murder rate was one of the highest in the U.S. at 59 murders per 100,000 residents, and politicians and police alike were criminal, corrupt, or incompetent. There was no question that New Orleans would have no lasting chance of survival without a comprehensive rehabilitation of the levees and drainage system. It was also necessary to strengthen the social cohesion of the population (Campanella 2006, 2007; Gelinas 2010). While many environmental scientists opposed reconstruction, most politicians advocated full reconstruction of the city. More important, however, was the question of whether the city would really be able

◻ **Table 5.4**   New Orleans: population and socioeconomic characteristics 2000 and 2010

	New Orleans	
	2000	2010
Inhabitants	484,674	343,829
Whiteness (%)	26.6	33.0
Blacks (%)	66.6	60.2
Inhabitants under 18 (%)	26.7	21.3
Average household income (US dollars)	43,176	37,176
People living in poverty (%)	27.9	24.4
Residential units	215,091	189,896
Thereof vacant	26,840	47,738

Sources: ▶ www.census.gov; Plyer (2011)

to recover from the disaster to its full extent (Kolb 2006; Vigdor 2008, p. 135). As this did not initially seem guaranteed, the aim was to save at least the world-famous French Quarter (◻ Fig. 5.36). Without the urban environment, the popular tourist quarter, whose disneyfication had long been criticised anyway in view of its artificial character, would finally have become a staged theme park (Souther 2007).

In retrospect, all the reconstruction plans of the following years seem chaotic, ill-considered, expensive and for the most part of little use. Repeatedly, foreign consultants were brought into the city, whose ideas or plans were usually discarded after a short time. New revitalization plans were developed again and again, but never realized. Interestingly, none of the new plans tied in with the plan that preceded them. As late as 2005, Mayor Ray Nagin formed a *Bring New Orleans Back Commission* consisting of planners, investors, and bankers who presented a first draft of how to rebuild the city in early 2006, but it was met with great resistance. Neighborhoods that had been entirely destroyed by Katrina were not to be rebuilt and replaced with green space. The plan split the city into two groups: The wealthy, who preferred to live in the higher elevations and whose homes had hardly been destroyed, were relatively satisfied, while the poorer, who lived mostly in the lower elevations, felt unwanted. The mayor abandoned this idea and suggested that the market, i.e. investors, should decide which parts of the city to revitalize. The laisser-faire policy met with little approval. In 2006, with financial support from the Rockefeller Foundation, another plan was developed and adopted in 2007 as the *Unified New Orleans Plan*. The entire city was to be rebuilt, although this was estimated to cost around $13 billion. The plan was generally welcomed, but was not financially feasible due to the high cost. Subsequently, the mayor developed the idea of recruiting international

◻ **Fig. 5.36**   Café in the French Quarter (March 2006)

star architects to erect eye-catching buildings along the Mississippi River in order to raise the city's global profile, which, as expected, was also criticized and soon discarded. Recognizing that spatial priorities needed to be set for rebuilding, a *Target Area Development Plan* was developed in 2007. This identified 17 subareas on which to focus revitalization efforts. The City Council approved this plan and appropriated $117 million in federal reconstruction funds for its implementation. Shortly thereafter, the new land use plan, New Orleans. *Blueprint for the Next 20 Years* was adopted (Comfort and Birkland 2010, pp. 670–673; Ford 2010). In addition, the *Road Home* program assisted returnees with $8.6 billion in federal funds. Owners of destroyed homes received an average of $66,000 each to rebuild. Even those who did not want to return home received aid from the program. Whatever was left of the house was bought by the government, and with the proceeds the owners paid any mortgages that still existed. The lots were put up for sale, with preference given to neighbors. If buyers took good care of the lots and possibly planted a garden, a portion of the purchase price could be refunded (Gelinas 2010, ▶ www.growinghomenola. org). The decision for or against New Orleans was facilitated and probably accelerated by the program. Those who returned made a conscious decision to stay in New Orleans and were willing to actively support the rebuilding of the city and the implementation of law and order. No small part of the reconstruction has been funded by nonprofit organizations, private donors, foundations, Hollywood stars, well-known entrepreneurs or politicians, and implemented with the support of famous or emerging architects. Special support was given to the construction of housing for the many homeless. Since a comprehensive plan was lacking for a long time, the sponsors were able to realize the houses or neighborhoods according to their own ideas. The results were sometimes bold designs that architecture critics describe as trend-setting. The actor Brad Pitt was able to win over the well-known architects Thom Mayne and Frank Gehry to design the houses he donated (Curtis 2009).

After the storm, houses were also demolished which, from the residents' point of view, could allegedly have been rehabilitated. Near downtown New Orleans, around 5000 socially disadvantaged families had been living in the Lafitte Projects and had been evacuated before the storm hit. The houses had been boarded up and had survived Hurricane Katrina relatively unscathed. The same was true for three other housing developments. The demolition of the *projects*, which were centers of drug trafficking and other criminal activity, had been demanded for years by politicians, who now took the opportunity to keep the houses barricaded and demolish them one by one. Residents were left homeless and no longer felt welcome in New Orleans (Ford 2010, p. 34). Meanwhile, the nonprofit Providence Community Housing is building new homes for different income groups on the site of the former Lafitte Project. While the anger of many residents over the demolition of the Lafitte Projects is understandable, large public housing developments have also been demolished in other cities such as Chicago and St. Louis in the past 20 years. The Lafitte Projects had been built during the New Deal era (1933–1936) and had spiraled downward over the decades, exacerbated by overcrowding (New York Times 12/26/2006).

5.8.3 Economy

The economy has been slow to recover from Hurricane Katrina in southern Louisiana. In New Orleans MA, the number of jobs fell from 61,000 to 519,000 in the first decade of the new millennium. Years after "Katrina", a portion of the population still suffers from post-traumatic stress and is unable to hold down regular employment. This is especially true for the poorer and poorly educated population, who cannot afford adequate medical care. These people are not available to the labour market. Many of those who have found jobs in the construction industry earn little. As a result, the average household income was lower in New Orleans in 2010 than it was 10 years earlier. At the same time, the need for well-trained skilled workers to expand the canal and lock system could not be met because the region's residents were not qualified to do so (Lyons 2011, pp. 293, 297; Plyer 2011; ▶ www.census.gov). After 2005, the economy was boosted by many incentives such as subsidies or tax reductions. However, a comparison of the economic sectors shows that only the construction and tourism and recreation sectors have far more employees today than in 2000. All other sectors have lost significantly (▶ www.bls.gov). New Orleans is again as popular a destination for tourists as it was before Hurricane Katrina. In 2011, 8.7 million visitors spent a record $5.47 million in the city on the Mississippi River (▶ https://eturbonews.com, 04/10/12). It seems that New Orleans is once again living up to its old image of the *Big Easy*. The city is promoting the expansion of digital, creative and green industries and the medical sector (Transition New Orleans Task Force 2010). Whether it will succeed in attracting forward-looking industries and sustainably changing the city's image remains to be seen.

5.8.4 Evaluation

Since 2005, things have indeed improved in New Orleans, while there are still major deficits in other places. Gradually, corrupt politicians have been replaced, and care has been taken to use federal and Louisiana state financial support responsibly. Since 2005, a great deal has been invested in schooling, and in 2012, 76% of all children attended a *charter school*. Students perform far better on tests today than they did a few years ago. Unfortunately, the crime rate has not been reduced. After temporarily dropping a bit, the murder rate in New Orleans returned to 2005 levels in 2012. While *"only"* five out of every 100,000 people are killed on average in the U.S., the number in New Orleans is 58. No other city in the U.S. has more people in prison than New Orleans. This is especially true for black males, one in seven of whom are in prison, have a criminal record, or are on probation (Gelinas 2010; Quigley and Finger 8/27/2012; ▶ www.bloomberg.com). On a positive note, the Army Corps of Engineers completed a 215 km rehabilitation of the lock, canal, and drainage system in Greater New

5

Orleans in 2012. An investment of $14.6 billion was made in order to be able to protect New Orleans from a flood of the century. Today, the city is said to have the best and safest system of its kind in the world (Zolkos 2012). Since Hurricane Katrina was certainly not the last violent storm to hit New Orleans, it is to be hoped that the new *Greater New Orleans Hurricane and Storm Damage Risk Reduction System* will effectively protect the city from damage.

5.9 Phoenix: A Paradise for Seniors?

Located in Arizona on the edge of the Sonoran Desert, Phoenix is one of the driest, warmest and sunniest regions in the U.S. The average temperature is 26 °C, and even in winter, temperatures rarely drop below 20 °C. In summer, temperatures of more than 40 °C are not uncommon. However, the dry climate is not as stressful as the hot and humid summers in the east and south of the USA. With a population of around 1.4 million, Phoenix is now one of the fastest growing cities in the USA, although settlement in the region began comparatively late. In fact, more than four million people live in Phoenix MA. By the fourteenth century, Hohokam Indians had settled in the region and had established irrigation canals and practiced agriculture (Keys et al. 2007, pp. 132–133). After the decline of Indian culture, the desert region did not attract renewed interest until the mid-nineteenth century, when a certain Wickenburg, on his way to California, happened to find gold here. Since the gold deposits did not prove profitable, Wickenburg rehabilitated the old canals and thus created the conditions for extensive irrigated agriculture. Toward the end of the Great War, when Egypt failed as a supplier of cotton, which was used to fill tires at the time, Goodyear, a tire manufacturer, bought the farmland to produce high-quality cotton. Due to the large amount of water used, the water table was rapidly declining and it was soon apparent that growing cotton at this location would not be profitable for long (Sun Cities Area Historical Society 2010, pp. 14–15). With the invention of the first primitive air conditioning systems in the 1930s, nothing stood in the way of living in Phoenix. The small desert town experienced its first boom during World War II when the American military located several training camps nearby. Air force bases were soon followed by industry. Beginning in the early 1940s, Goodyear Aircraft manufactured in nearby Litchfield, and AiResearch opened a plant at Sky Harbor Airport, Phoenix's airport. Other plants, such as the Allison Steel Company, founded in Phoenix in the 1920s, specialized in the production of war equipment. The region's economic boom continued after 1945, and Phoenix developed into a major center

for the aircraft and electronics industries. By the early 1950s, more than 90% of all homes had air conditioners, many of which were manufactured in Phoenix. Because production costs were low due to a lack of union influence and low taxes, Phoenix was able to develop into an important industrial center. In addition, the city benefited from its isolated location and was able to move its parish boundaries further and further outward to make room for its growing population. People moved to the region from all over the country to take the well-paying jobs. Industrial development also created a demand for financial services (Konig 1982, pp. 19–22, 27–33). However, only four of the 500 largest companies have their headquarters in the City of Phoenix and two others in neighboring Tempe. None of the companies are among the 100 largest in the nation by revenue. US Airways (ranked 208) and Petsmart (ranked 400) are probably the best known (▶ www.fortunemagazine. com).

5.9.1 Sun City

The Phoenix MA is as highly fragmented socioeconomically and ethnically as other metropolitan areas in the U.S. (◘ Table 5.5, ◘ Fig. 5.37). In the City of Phoenix and Tempe, the median household income is just below that of the State of Arizona, but far fewer Hispanics live in the University City of Tempe than in Phoenix. Scottsdale is a preferred place to live and features a very high household income. It is also home to more than twice as many seniors as Tempe or the state of Arizona. Most strikingly, in Sun City and in Sun Lakes, more than 70% of residents are at least 65 years old. In Sun City West, this is even true for more than 80% of the population. At the same time, these communities are almost exclusively white and have decidedly few Hispanics. The large number of seniors in some Phoenix MA communities is not a coincidence, of course, but is the result of planned development. When developer Del E. Webb was looking for a site to build a senior housing development in the late 1950s, the former cotton plantation located nearly 20 km northwest of Phoenix seemed to be the ideal location because water rights were available for a large tract of land (Sun Cities Area Historical Society 2010, pp. 14–15). Sun City was not the first settlement for the elderly in Arizona. Nearby, Youngtown had already been laid out in 1958 to serve the same demographic (Blechman 2008, p. 29). While Youngtown's investors had limited themselves to building housing suitable for seniors, Sun City was intended to offer all the requirements for an active retirement and was marketed as "*An Active Way of Life.*" Also new was Sun City's circular layout, punctuated by green spaces

Table 5.5 Structure of selected municipalities in the MA Phoenix 2010

Community	Inhabit-ants	White (%)	Black (%)	Hispanics (%)	Median household income (US dollars)	64 years and older (%)
Phoenix	1,444,656	65.9	6.5	40.8	48,596	8.4
Tempe	164,268	72.6	5.9	21.1	48,618	8.5
Scottsdale	221,020	89.3	1.7	8.8	71,816	20.0
Sun City[a]	37,499	96.5	1.4	2.8	36,117	74.9
Sun City West[a]	24,535	97.8	0.8	1.2	34,252	83.6
Sun Lakes[a]	13,975	97.0	1.2	1.8	39,142	72.9
Arizona	6,392,015	65.9	6.5	30.1	50,752	14.2

Source: ▶ www.census.gov
[a]CPD = census designated area

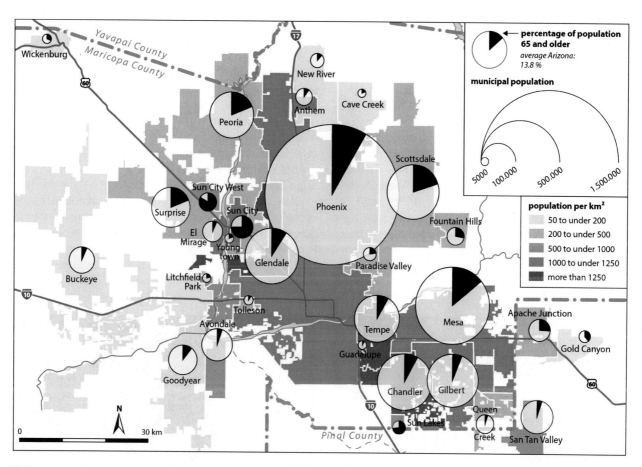

Fig. 5.37 *Phoenix metropolitan area.* (Data and map basis ▶ www.census.gov)

and small waterways. A range of sports, leisure and shopping facilities tailored to older people were to be available by the time the first residents moved in. Since a diverse offer could only be profitable in the long run for a large number of users, the construction of more than 16,000 residential buildings and apartments for around 32,000 residents on 67 km² was planned. Although many critics doubted that so many people would want to spend their twilight years only among older people in a neighborhood that was homogeneous in every respect,

5

the demand exceeded the investors' wildest expectations. In September 1959, when the first model homes were open for viewing and the sales office opened, 400 future residents put down a $500 deposit on a home. In early January 1960, Sun City's grand opening was celebrated with a 3-day festival attended by more than 100,000 onlookers. On those days, 237 homes were sold (Sun Cities Area Historical Society 2010, pp. 16–25).

Despite the great initial interest in Sun City, sales figures soon began to decline. In the mid-1960s, the investors even had to fear for the lasting success of the senior citizens' settlement. As part of the second construction phase, a leisure center had been built exclusively for the new residents. The displeasure of Sun City's "*pioneers*" had become common knowledge and had had a negative impact on the image. Moreover, investors had mistakenly believed that the cheapest possible houses along the lines of Levittown were particularly desirable. In fact, Sun City was the first to offer homes for less than $10,000 that were right on a golf course. However, seniors preferred better homes and were willing to pay for them. Based on this realization, it was decided to build higher quality homes in the future and to make all community amenities in each subdivision of Sun City equally accessible to all residents. After selling only 400 homes in 1965, about 1800 homes were again occupied in 1967. Over the years, better and better homes were built, some with their own swimming pools (Sun Cities Area Historical Society 2010). The high demand led to the addition of several other circular housing developments to Sun City, and finally to the construction of the Sun City West development east of Sun City between the late 1970s and 1998. Parallel streets, laid out in the shape of a semicircle, form the layout of Sun City West. The houses in the two *retirement communities* are standardized and can be selected by catalogue. They are always single-storey bungalows painted in light colours (◘ Fig. 5.38). The houses are

complemented by a double garage with direct access to the kitchens. There are a few cacti in the small front gardens as there are in the rear gardens. The gardens are otherwise devoid of vegetation and the ground is covered with gravel. Overall, the houses and very wide streets give a monotonous impression but are low maintenance and laid out to suit the elderly (author's observation). In early 2013, the average purchase price for a home in Sun City was $107,600 and in Sun City West $172,600 (► www.zillow.com).

In 1962, residents of the *retirement community* formed the Sun City Home Owners Association (SCHOA) and took the fate of the community into their own hands. Their first goal was to establish health insurance, which they developed with Continental Casualty Company exclusively for Sun City residents. A year later, Sun City's senior citizens voted by referendum to oppose political autonomy, i.e., to become an independent city. Sun City is governed by Maricopa County (population 3.8 million), but represents only a fraction of the population. The retirement community does not have its own mayor, administration or police force, but still performs all municipal functions on its own. Organizationally, this is done through Recreation Centers of Sun City, Inc. (RCSC), and SCHOA, which consolidates and organizes all activities and planning. Sun City and its sister community, Sun City West, have eight golf courses, several tennis clubs and swimming pools, recreation centers for the pursuit of every conceivable hobby, and several shopping centers (◘ Fig. 5.39). About 60% of Sun City residents volunteer according to their abilities in self-government or in the numerous recreational clubs in the settlement. Even the hospitals of Sun City and Sun City West are supported by about 1500 volunteers. Resident involvement enables a wide range of services to be offered in all areas, saves costs and promotes a sense of community. Sun City is proud of the dedication of

◘ **Fig. 5.38** Bungalow in Sun City

◘ **Fig. 5.39** Sun City leisure center

its residents and likes to refer to itself as "*The City of Volunteers*" (Sun Cities Area Historical Society 2010, pp. 40–41, 69).

5.9.2 Sun City as a Role Model

Around the world, similar settlements for an older target group have been built along the lines of Sun City. For real estate developers, the construction of senior housing estates is an interesting line of business with very good prospects for the future, because in view of the aging population in the USA as well, the number of potential buyers is rising rapidly. Since 1960, Webb Corp, which was purchased by Pulte Homes in 2001, has sold about 80,000 homes and apartments to active seniors in more than 50 senior housing developments in 20 states. The American Seniors Housing Association (2012, p. A8) reports that the 50 largest companies specializing in senior-friendly homes alone own just over 500,000 housing units, although only some of these are in developments designed specifically for this age group. Very different data are available on the number of senior housing developments in the USA. The homepage ▶ www.55communityguide.com presents 701 *retirement communities* across the country, of which 50 are in Arizona and 71 in Florida, while Blechman (2008, p. 220) speaks of an estimated 1300 settlements of various sizes specifically for older residents, some of which, however, were still under construction or expansion. In more recent times, however, more than a few seniors prefer to move into age-restricted housing in downtowns or in locations close to downtowns, as they do not want to miss the amenities of cities and probably do not want to live only among their peers. Also unattractive are older *retirement communities* where most residents use walkers, which Blechman (2008, p. 7) refers to as *gerotopia*. *Retirement communities* and new residents prefer active adults or young old people who are fit to pursue many hobbies and who are not dependent on care or nursing.

The featured *retirement communities* in Phoenix MA are open access, while many other communities of this type are surrounded by fences or walls and access is controlled. Probably the largest *gated retirement community* is The Villages near Orlando in central Florida. The Villages is divided into dozens of gated subdivisions and spans three counties and about 80 square miles. Each subdivision *is* a separate entity that is part of the whole. The Villages include two man-made downtowns and several shopping centers. The Villages are served by a 160 km trail system for electric-powered golf carts on which residents travel all distances. At the beginning of the new millennium, about 75,000 people lived here, and expansion to about 110,000 residents was planned. When completed, The Villages will be almost as big as Manhattan. There will be dozens of pools, hundreds of clubs for a variety of hobbies, and more than three dozen golf courses. The biggest advantage considered by many residents is the fact that there are no children living in The Villages (Blechman 2008, pp. 4–5, 39).

5.9.3 Criticism

No ethnic or socio-economic group in the U.S. lives as segregated as seniors in *retirement communities*, which are highly controversial, although it must be acknowledged that this is voluntary segregation (◘ Fig. 5.40). Proponents of retirement communities argue that older people have different needs than younger people, while opponents believe that younger people and families are discriminated against. Residents are generally required to be at least 55 years old. Repeatedly, court cases have arisen because grandparents have taken in a minor grandson or because widowed residents married a younger woman and did not want to move out. The courts have usually ruled against the younger residents because age discrimination is allowed in the United States. The federal *Fair Housing Act of* 1968, while prohibiting discrimination on the basis of ethnicity, does not include provisions to discriminate

◘ **Fig. 5.40** Assisted living in San Francisco

against or favor certain age groups. In 1988, the courts confirmed that renting or selling apartments or houses exclusively to the elderly was not prohibited by law. It was also permissible to prevent children from residing in these settlements (Blechman 2008, pp. 66–79). In 1995, Congress even legislated housing rights exclusively for residents 55 or 62 years of age and older with the passage of the *Housing for Older Persons Act* (Lynn and Wang 2008, p. 37). Arguably the biggest problem is that seniors are limited in their support of American society because their taxes are not used to maintain schools. This is objected to by many critics as being extremely selfish. Different states have different rules about whether and to what extent seniors must pay for facilities they do not use. Many seniors choose where to live in the future based on the amount of taxes they pay, and often choose Florida, where it is clearly stated that they do not have to pay for schools. This is considered unfair because it puts the burden of taxes solely on families with children (Blechman 2008, pp. 222–225).

5.10 Seattle: High-Tech Location on the Pacific Ocean

With a population of 620,000, Seattle is the largest city in the northwest of the USA, located on the 150 km long and widely ramified Puget Sound. The Seattle MA, which includes the three counties *of* King, Snohimish and Pierce with the cities of Seattle, Bellevue, Everett and Tacoma, is home to 3.5 million people, almost half the population of Washington State, but accounts for 69% of the state's economic output. Based on commuting patterns, the Puget Sound region was also designated, with a total population of 4.3 million (▶ www.census.gov; Katz 2009). Hardly any other city can boast as many epithets as Seattle, which has been dubbed *Jet City, Competitive Global City, Global Justice City, Curative Philantrophy City, Emerald City* and, more recently, *Music City*. In addition, Seattle is one of the most important high-tech locations in the United States. In 2011, Scientific American magazine awarded Seattle first place in the categories of *most tech-friendly* (2nd place: San Francisco Bay Area, 3rd place: Los Angeles), *best internet access* (2nd place: Atlanta, 3rd place: Washington, D.C.) and *overall technology performance* (2nd place: Orlando, 3rd place: Washington, D.C.). Seattle is also regularly ranked highly in other rankings (Bushwick 2011).

Seattle is not only about 4800 km away from Boston, New York or Washington, D.C., but also from the economic centers of the American West Coast San Francisco and Los Angeles about 1300 and 1800 km respectively, but the isolated location has proven to be advantageous. Seattle is now considered the capital

■ **Fig. 5.41**   Close integration of city and water in Seattle

of the Pacific Northwest of the United States, which includes the states of Washington, Oregon, Alaska, Idaho, and western Montana. These states represent Seattle's hinterland, as they are served in many ways by the city (Brown and Morrill 2011, pp. 5–6, 35). The positive development of the Puget Sound region is due to an interplay of endogenous and exogenous factors. The mild climate, abundant forests, and deep natural harbor have had a beneficial effect (■ Fig. 5.41). In addition, successful entrepreneurs have repeatedly located in the region, providing an early foundation for an innovative milieu that has produced other major talents, and the U.S. defense industry has made substantial investments in the site. More recently, the aircraft industry has been supported by large subsidies. The Puget Sound region (■ Fig. 5.42) has several very good universities and the region's residents are better than average educated. While only 28.2% of all U.S. residents have at least a bachelor's degree, this is true for 55.8% of Seattle residents. Household income is also well above the U.S. average, but correlates with high housing prices and cost of living. However, incomes are comparatively evenly distributed and child poverty is low. As a high-tech location and innovation pool, Seattle is comparable to Boston or the Research Triangle in North Carolina, but does not have Boston's ever-simmering racial problems and poor neighborhoods, and is much more compact than rural North Carolina (▶ www.census.gov; Katz and Jackson 2005).

5.10.1 Rise to a High-Tech Region

George Vancouver had laid out in 1792 in the wide-branching bay, which is about 150 km long, but the first permanent settlement did not come into being until 1851. The first sawmill was opened a year later. This

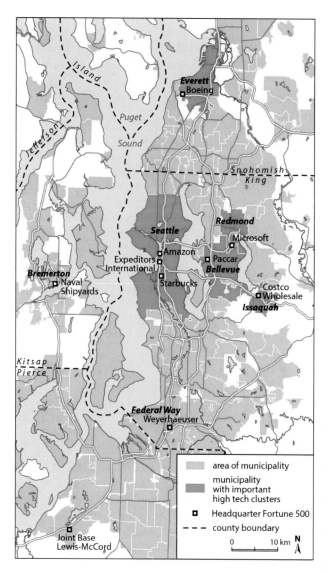

☐ **Fig. 5.42** Puget Sound region. (Map base ▶ www.census.gov)

produced lumber for San Francisco, which was growing rapidly due to the California gold rush. In the late 1860s, the first railroads were built, bringing lumber to the sawmills from greater distances. Connection to the transcontinental railroad came in 1883, and lumber from the heavily forested Northwest was now transported by rail to much of the North American continent. In 1897, a ship loaded with gold coming from Alaska docked in Seattle. Immediately, many potential gold miners migrated to Seattle to make their way from there to Alaska. Most, however, stayed in the Puget Sound region and worked in the lumber industry, farming, fishing, or in the nearby coal mines. During the Great War, Bill Boeing and Conrad Westervelt formed the Boeing Company and contracted with the Navy to build airplanes. Unable to build even one airworthy device during the war, the Boeing Company produced furniture for

a time but remained busy developing airplanes. In the 1920s, Boeing's first airplanes carried mail at first, but by 1929 they also carried passengers. In 1932 Mac McGee founded Alaska Airlines in Seattle with the aim of supplying Alaska from here. Today, the hub and headquarters of Alaska Airlines are still located in Seattle (Brown and Morrill 2011, pp. 19–21, 38).

During World War II, further foundations were laid for the region's later economic success. As a manufacturer of aircraft for the Air Force, the Boeing Company's workforce grew from 4000 in 1940 to 50,000 just 4 years later. Bremerton Naval Shipyard also played an important role, employing as many as 32,500 workers in ship repair. Employment at Boeing and the shipyard plummeted after the war ended, but the Korean War (1950–1953) and the Vietnam War (1965–1973) rebounded military investment in the Puget Sound region, and the Boeing Company soon became the world's largest manufacturer of civilian and military aircraft. Today, Bremerton Naval Shipyard is one of the largest repair yards for American warships. The bay also serves as a base for nuclear submarines and is homeport to several aircraft carriers, and south of Tacoma are Fort Lewis Army Base and McChord Air Force Base, which merged in 2010 to form Joint Base Lewis-McChord with about 25,000 military and civilian employees (Brown and Morrill 2011, pp. 21, 27). Over the decades, the military's investments have brought many well-trained professionals to the region, who have been critical to its rise as one of the leading high-tech regions in the United States. The boom in commercial air travel in the 1960s benefited the Boeing Company, and by extension the city of Seattle, extraordinarily. At times Seattle was almost as dependent on the aircraft industry as Detroit was on the automobile industry or Baltimore with Bethlehem Steel on the steel industry and could be described as a modern company town. When Boeing was doing poorly, Seattle suffered as well. In 1969, Boeing plants in the region peaked at 105,000 employees. During the recession of the early 1970s, Boeing laid off more than 60% of its workforce, plunging Seattle into a severe crisis. A significant recovery was not felt until the early 1980s. At the same time, the high-tech sector began to diversify. While in the mid-1970s half of the people employed in this sector were still working in the aircraft industry, today this is only true of just under a quarter. The high-tech sector, which accounts for 40% of all jobs in the Seattle MA and is characterized by high investment and high value added, employs not only engineers but also many computer professionals. The Puget Sound region is nowhere near as dependent on the aircraft industry as it was a few decades ago. Nevertheless, the local economy experienced several downturns beginning in 2000, caused by the decline in high-tech stocks (*dot-com crash*), the 9/11 attacks in 2001, the SARS epidemic in 2002/2003, and the mortgage and banking crisis in 2008 (Brown and Morrill 2011, pp. 23–24).

The Boeing Company moved its corporate headquarters to Chicago in 2001, which was, of course, heavily criticized on the West Coast. Nevertheless, the relocation was more of a symbolic act with little impact on the Puget Sound region, as production of the huge aircraft remained in Everett and few jobs were lost in the region (Brown and Morrill 2011, pp. 25–27). Things only became dangerous when Boeing announced in the spring of 2003 that it would build the new "*Dreamliner*" outside of Washington State. Several states, such as Michigan, Texas, and California, tried to lure Boeing with enormous subsidies. In the Puget Sound region, 800–1000 jobs at Boeing plants and another 17,000 or so jobs at suppliers were at risk. Washington State countered by promising to pay the aircraft manufacturer $3.2 billion over a 20-year period. If the "*Dreamliner*" were to be built in the Seattle region, Boeing was to receive $160,000 per year per employee, although average annual salaries were only $65,000. Boeing accepted the offer and built a new plant in Everett. The subsidies are controversial not only because of their large size, but also because the individual parts of the "*Dreamliner*" are produced at various locations in Japan, Europe and the USA, flown to Everett and "only" assembled there. Meanwhile, the "*Dreamliner*" is additionally assembled in South Carolina, probably because labor costs and union influence are lower there than in Washington State (Kavage 2004; ▶ www.boeing.com). Since the planes are delivered to airlines around the world anyway, and the cities are in strong competition with each other not only on a national level but also on an international level, final assembly could theoretically take place at any other location in the world. *Jet City*'s future is by no means assured, especially since Airbus, a serious competitor, has emerged in Europe in recent years. New troubles announced themselves in early 2013 when the new "Dreamliner" was grounded until further notice due to problems with the battery, which can lead to fires, and other glitches. Immediately, a wave of layoffs began at Boeing and supplier companies in the Puget Sound region.

5.10.2 Diversification

In view of the uncertain outlook for the aircraft industry, it is favorable that other globally active companies with good future prospects have developed in the Puget Sound region in recent decades (◻ Table 5.6). In 1975, Paul Allen and Bill Gates, both originally from Seattle, co-founded Microsoft in New Mexico to develop programs for the "*Altair 8800*" computer produced there. In

◻ **Table 5.6** Companies in the Puget Sound region in 2012 that are among the top 500 in the country by revenue, according to Fortune, the U.S. business magazine

Rank (turnover)	Company	Business segment	Location
24	Costco Wholesale	Wholesale	Issaquah
37	Microsoft	Software and hardware manufacturers	Redmond
56	Amazon.com	e-commerce mail order company	Seattle
159	Paccar	Truck manufacturer	Bellevue
227	Starbucks	Café chain	Seattle
374	Weyerhaeuser	Forestry	Federal Way
395	Expeditors International of Washington	Logistics	Seattle

Source: ▶ www.fortunemagazine.com

1978, they moved Microsoft, spelled without a hyphen from then on, to Bellevue near the city of Seattle, where a few years later they developed the operating system MS-DOS (Microsoft Disk Operating System) on behalf of IBM, without which soon hardly any personal computer worked. At the end of 2012, Microsoft had 97,000 employees worldwide, just over 41,000 of them in Seattle MA (▶ www.microsoft.com). Amazon, which was founded in Seattle in 1994 by Jeff Bezos and initially sold only books, but later increasingly all kinds of other products over the Internet, has also quickly developed into a globally active company. This is also true for the Starbucks Coffee Company. The first Starbucks was opened in 1971 in Pike Place Market on the edge of downtown Seattle by three professors (◻ Fig. 5.43). The upswing to a world-wide chain of coffee houses began 10 years later, when the New Yorker Howard Schultz became co-owner and successfully transformed Starbucks along the lines of Italian coffee bars. Although this has not really succeeded from a European point of view, as the takeaway coffee in cardboard cups is little reminiscent of Italian quality of life, the licensed and group-owned coffee houses have now been able to establish themselves in more than 50 countries due to very good marketing. The company is still headquartered in Seattle (▶ www.starbucks.com). In addition to the companies mentioned above, there are

Fig. 5.43 One of the first Starbucks in Seattle

hundreds, if not thousands, of small and micro companies in the high-tech industry that are suppliers or *spin-offs* of the giants. The region has also been positively impacted by the high salaries and bonuses common in this sector, which have gone to many charitable projects or donations to museums or the region's universities (Brown and Morrill 2011, pp. 23–25, 38–39). Overall, the Seattle MA is home to seven of the 500 largest companies in the United States by revenue, including three in the City of Seattle (▶ www.fortunemagazine.com). Seattle is not one of the top banking centers in the U.S., but it is still the financial center of the region. The ports of Tacoma and Seattle are, after those of Los Angeles and Long Beach, the busiest ports on the Pacific coast of the United States in terms of cargo handling. Since almost all the imports and exports of the northwestern United States are handled here, the ports are considered the gateway to Asia. China is the most important trading partner, accounting for almost 50% of transshipments (excluding Hong Kong, Taiwan and Macau). The highest concentration of jobs is in downtown Seattle and in a corridor extending north to the University of Washington. There are also many jobs in Bellevue and neighboring Redmond, which is home to Microsoft and the U.S. headquarters of Japanese manufacturer Nintendo. There are several high-tech clusters in Tacoma, Everett, and Bremeton, which are dominated by the aircraft industry and defense manufacturing facilities (Brown and Morrill 2011, pp. 5–6, 28, 43).

The terms *Competitive Global City*, *Global Justice City* or *Curative Philantrophy City* for Seattle are only contradictory and inconsistent at first glance. Since globally active companies such as Microsoft, Amazon and Starbucks have to compete globally against competitors in the USA and other countries, Seattle MA is indeed a *Competitive Global City*. *Global Justice City* harkens back to the street fighting in 1999 when protesters violently lobbied for the dissolution of the World

Trade Organization (WTO) and fairer treatment of less developed countries under global trade agreements. *Curative Philantrophy City* recalls the high level of giving by many foundations in the region, of which the Bill and Melinda Gates Foundation is the best known. The world's richest foundation is committed to the development of medicines and is particularly keen to help the poor in Africa (Brown and Morrill 2011, pp. 49–50).

5.10.3 Outlook

The ongoing settlement of Puget Sounds has greatly altered the region's ecosystem, negatively impacting forest, fish and wildlife populations. Hills have been leveled, waterways channeled, and marshes drained, not only impeding salmon from reaching their spawning grounds, but also displacing the population that once lived here. Echoing the *Emerald City* that transforms from diamond-studded to less attractive in the *Land of Oz* fable, Klingle (2006) also refers to Seattle as the *Emerald City*. However, the region was able to benefit early on from the environmental awareness and commitment of local politicians. Senators Warren G. Magnuson and Henry M. Jackson, both natives of Seattle MA, lobbied Washington, D.C., in the 1960s and 1970s for passage of the *Coastal Management Act*, the *National Environmental Policy Act*, the *Wilderness Act*, and the *Clean Air and Clean Water Act*. Spurred by the two senators' efforts, other conservation initiatives such as the North Cascades Conservation Council were established in Washington State (Brown and Morrill 2011, p. 22). Seattle works closely with the Canadian city of Vancouver, located to the north, and Portland, located to the south in the US state of Oregon. Citizen and non-governmental engagement with the region is strong and has led to the Puget Sound region, centered on Seattle, becoming known as particularly environmentally conscious or green (Ott 2001).

Although attempts were made early on to counteract increasing urban sprawl, the core city of Seattle is nevertheless losing importance in relative terms. At the same time, the surrounding area is experiencing major growth, as the population in the *outer suburbs* and *exurban areas* has increased far more than in the city of Seattle and some *inner suburbs* in recent decades. The inner suburbs lost about 30,000 jobs from 2000 to 2009 alone, while simultaneously adding 90,000 jobs in the suburban area (Cox 2010, 2011d). However, it has a comparatively large retail offer. Most attractive is the historic Pike Street Market, which was renovated in the 1980s and is popular with locals and tourists alike. A major problem, on the other hand, is the Alaskan Way Viaduct, which closes off two levels of downtown from the waterfront (◘ Fig. 5.44). After the viaduct was damaged by an earthquake in 2001, it was decided in 2009, after much discussion, to move the road into a tunnel similar to the one in Boston, at a cost of $4.2 billion (◘ Fig. 5.45) (Lind 2009).

◘ **Fig. 5.44** Elevated Alaskan Way Viaduct

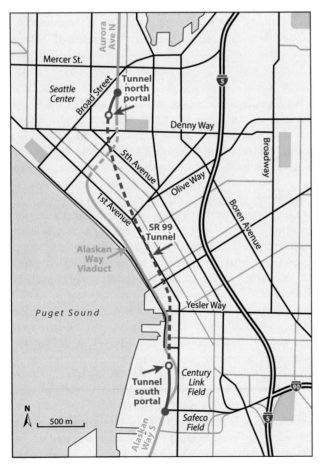

◘ **Fig. 5.45** Tunneling in Seattle. (Adapted from City of Seattle records)

In 2008, the Seattle MA suffered greatly from the recession and the unemployment rate was in some cases far above the national average. Recovery was expected soon, as there had been many start-ups in nanotechnology, biomedicine, and healthcare services in the preceding years. In addition, the region was considered a leader in building eco-friendly homes, promoting electric vehicles, and sustainable development. Thus, there was no shortage of forward-looking industries (Katz 2009). In fact, however, it was not until 2012 that a noticeable upturn occurred and the unemployment rate fell to 6.9% at the end of the year (US: 7.8%) (► www.bls.gov). Quite unexpectedly, the secondary sector has made a major contribution to the recovery of the economy. The U.S. has benefited in recent years from rising wages and the resulting decline in Chinese competitiveness and declining manufacturing in Europe, and from the beginning of 2010 through March, about 470,000 new industrial jobs were created. 70% of all research and development (R&D) spending in the U.S. is by industrial companies. In the Seattle-Bellevue-Everett region, the figure is even much higher, with half of all R&D spending by aircraft manufacturer Boeing. In a 2012 ranking that compared secondary sector growth in 100 locations, Seattle-Bellevue-Everett received first place, ahead of Oklahoma City and Salt Lake City (Kotkin 2012). Overall, it appears that the future outlook for the Puget Sound region is very bright. There continues to be a lot of emphasis on the high-tech sector, but at the same time the city is working on a new image. In 2008, Mayor Greg Nichols announced that Seattle would market itself as the *City of Music*. Indeed, Seattle's fame as a major music hub dates back to the 1960s and is closely associated with the names of Jimi Hendrix, Ray Charles and Quincy Jones. In Seattle, more than 11,000 people are employed in the music industry in the narrow sense and even more than 22,000 in the broader sense. In total, the industry generates about $2.6 billion in revenue annually. Seattle hopes to soon be mentioned in the same breath as Nashville or Austin, which have long been of great importance for music (Baumgarten 2009).

5.11 Las Vegas Between Hyperreality and Bitter Reality

Located in a high valley of the Mojave Desert in Nevada, Las Vegas was a settlement area for the Paiute Indians due to natural springs. In 1829 traders from New Mexico discovered the location, in 1855 Mormons temporarily settled in the region, and in the 1860s the American Army established a base here with Fort Baker. More significant to the growth of the settlement, however, as is so often the case in the United States, was its connection to the railroad, which linked the settlement to Salt

Lake City and Los Angeles shortly after the turn of the twentieth century. In 1905, Las Vegas was elevated to the status of a city (Schmid 2009, p. 92). Las Vegas owes its unrestrained growth to gambling, which includes all kinds of lotteries, horse and dog betting, and casinos. Since there has never been any significant industry in Nevada, the state has not suffered the consequences of deindustrialization, but has immediately developed into a service location. However, American attitudes towards gambling have always been ambivalent, with the moralistic Puritans condemning it while others saw it as a harmless pastime. In the West, California had long been the stronghold of gambling, which was controlled there by Chinese who had followed the gold rush of the mid-nineteenth century. As crime and corruption associated with gambling increased, it was banned outright in California in 1891, whereupon it moved to neighboring Nevada, but was also banned there in 1910 (Dunstan 1997). During the recession triggered by the stock market crash in 1929, individual states discovered gambling as a tax revenue and allowed it temporarily; in Nevada it was legalized permanently in 1933. Near Las Vegas, the Boulder Dam (now Hoover Dam) was under construction, and it was hoped that gambling would attract tourists. It was also hoped that gambling, which had been illegal until then, could be better controlled, as it was largely in the hands of criminal gamblers (*mobsters*). The growth of Las Vegas also accelerated as Los Angeles sought to rid the city of prostitution and gambling in the late 1930s; both shifted to the desert city. Henceforth, gambling, an often questionable entertainment industry, crime, and *mobsters* characterized the city (Jayne 2006, p. 69). Particularly notorious was Benjamin "Bugsy" Siegel, who moved from California and was granted a license in Las Vegas despite his bad reputation. In 1946, he opened the Flamingo Hotel, the first modern theme hotel on the Las Vegas Strip. Since gambling was still banned in the other states, Las Vegas extended its supremacy undisturbed for several decades. In the 1950s, the legalization of striptease performances and fast-paced weddings and divorces also increased its appeal to certain visitors (Rothman 2002; Rothman and Davis 2002). At the same time, the influence of the U.S. Cosa Nostra, which controlled numerous hotels, increased. Portions of casino profits were siphoned off by *mobsters* before they could be taxed, ending up with family bosses who controlled casinos far from Las Vegas in cities such as Chicago and Miami (Jayne 2006, p. 69). In 1976, first New Jersey and soon other states legalized gambling. Faced with increasing competition, Las Vegas tried harder to lose the image of the "*sinful*" city and attract not only gamblers but also other visitors. The plan worked. Today, Las Vegas is probably the most famous casino location in the world, and the dream worlds of the Strip enchant young and old. But all that glitters is not gold in this gambler's paradise (Hahn 2005a).

Few visitors to the world-famous Strip, with its many hotels and casinos, are likely to know that they are not in Las Vegas, but in the township of Paradise just outside the city, which is also home to McCarran International Airport, the University of Nevada and the Convention Center. Las Vegas describes the actual city, the Strip, or the entire metropolitan area and has become a brand name used by the Las Vegas Convention and Visitors bureau to market the region (◻ Fig. 5.46). In the following, the name Las Vegas is also used to refer to the entire metropolitan area, which fills almost the entire valley between several mountain ranges at an elevation of about 600 m (◻ Fig. 5.47). Greater Las Vegas is identical to Clark County, which includes, in addition to the core city of Las Vegas with 580,000 residents, the much smaller cities of North Las Vegas, Boulder City, Henderson, Mesquite, and large uncooperated areas such as the township of Paradise. The region's population has increased by a factor of 30 since 1940 to nearly two million in 2010, an exceptionally high figure even for the United States. The population has grown particularly rapidly since 1990, when Greater Las Vegas

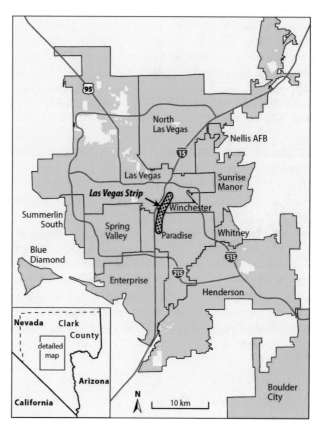

◻ **Fig. 5.46** Greater Las Vegas. (Map basis ▶ www.census.gov)

5

Fig. 5.47   View over Greater Las Vegas

Fig. 5.48   Hotel New York-New York on the Las Vegas Strip

had only 770,000 residents (▶ www.census.gov). In the 1990s, 4000–6000 people per month moved to Las Vegas in some cases. Water for the desert city is drawn from the Colorado River without regard to the declining water levels of Lake Mead, a reservoir some 50 km away. Tucson, AZ, which has a similar climate to Las Vegas, uses 600 l of water per day per resident, but Las Vegas uses nearly 2300 l. Las Vegas is part of the *new sunbelt*, to which people move not only from the *rustbelt*, but also from the *old sunbelt*, which includes California, Texas or Florida. More people move from California, where most of the visitors to the casinos on the Strip come from, than from any other state. In addition, Las Vages exerts a great attraction on Hispanic immigrants. From 1980 to 2000, the Hispanic population in the area increased by 900% to around 350,000–4,00,000. Greater Las Vegas is as segregated as other regions of the United States. Affluent and predominantly white communities contrast with very poor communities with a high proportion of Hispanics (Ventura 2003, ▶ www. census.gov). Residential development has largely been by private developers who prefer to create *gated communities*. The retreat into gated communities reveals in some ways a schizophrenic behavior of residents, many of whom on the one hand enjoy visiting the city's sometimes semi-seedy entertainment venues and casinos, but on the other hand entrench themselves behind walls or fences. Although no exact figures are available, the proportion of people living in *gated communities* is said to be nowhere higher in the USA than in Las Vegas (Ventura 2003, p. 103).

5.11.1   Hyperreal Worlds

Opened in 1946, the Flamingo Hotel ushered in a new era. With its spacious bathing area, it offered the oppor-

tunity to combine gambling with a relaxing holiday. The modern *resort hotel* was born. Since the 1980s, the idea has gone from strength to strength. All new large Las Vegas hotels have a theme, choosing exotic names from an American perspective such as Excalibur, Bellacio or Luxor, and cities with a positive image such as New York, Paris or Venice (Gottdiener 2001, pp. 105–106). The New York-New York hotel, part of the Metro Goldwyn Mayer (MGM) group, markets itself as *The Greatest City* in *Las Vegas* (▣ Fig. 5.48). Opened in 1997, the $460 million complex is composed of scaled-down but faithful replicas of New York skyscrapers such as the Chrysler Building and the Empire State Building, with more than 2000 hotel rooms. It also features replicas of Ellis Island, Grand Central Station, the Brooklyn Bridge and the Statue of Liberty. The leisure facilities include several large themed restaurants, a roller coaster and a swimming area. The almost 8000 m^2 large foyer of the hotel is thematically divided into different areas such as "*Park Avenue*", "*Times Square*" or "*Central Park*". The arriving guests have to find their way to the reception through more than 2000 one-armed bandits. In addition, there are several hundred gaming tables for roulette, black jack or baccarat (Hahn 2005a, p. 25; Onboard Media 2002).

According to postmodern theorists, the boundary between reality and fantasy can no longer be clearly defined. Las Vegas rises from the desert from afar and advertises itself. The Strip represents hyperreality at its most perfect, for the presentation is more real than the reality depicted. Las Vegas does not depict the real America, but its simulation. The simulated worlds have become the authentic America, staged, clean and controlled (Baudrillard 1993, pp. 91–93, 1994). The extent to which the hyperreal Las Vegas has displaced the real America was made clear when the United States Postal Service issued a stamp in 2011 commemorating the 125th anniversary of the birth of "*Lady Liberty*" in New York Harbor. It was based on the Statue of Liberty at the New York-New York Hotel in Las Vegas, which is only 14 years old, and its proportions are not true to the original. Only an attentive collector noticed the mistake (New York Times 14/4/2011).

Every new hotel has to be bigger and more exciting than any of its predecessors. Some hotels, such as Caesars Palace and the Venetian, house entire shopping malls where the most expensive designers can be found. Many amusements are free and quite suitable for families, as the Strip itself has become a stage with many attractions such as pirate battles or the huge fountains of the Bellagio hotel. The *resort hotels* have huge pool areas, offer roller coasters and IMAX cinemas, so there is never a dull moment. There are still the topless bars, but they are well hidden off the Strip. Numerous shows feature world-renowned artists performing every night, often for years, in perfectly staged performances. The cost of building the theater and producing Cirque du Soleil's spectacular show "Kà," which has been shown daily at the MGM Grand Hotel since 2004, is said to have cost more than $150 million. Gambling, entertainment and consumption have merged to an extent that can hardly be surpassed, which is why the *resort hotels* are described as cathedrals of consumption. Postmodern consumer landscapes of the highest perfection have been created that never cease to amaze visitors (and drain their money). Moreover, Las Vegas has succeeded in becoming a magnet for cultural travelers by means of several high-profile museums such as the Wynn Collection of Fine Art (Gottdiener 2001, pp. 105–116; Gottdiener et al. 1999, p. 38; Hahn 2005a; Ritzer and Stillman 2004, pp. 83–99; Schmid 2009, pp. 164–178).

Today, hotels and casinos are no longer built and operated by criminals but by global corporations, and investors often come from other parts of the world. However, the globalisation of the hotel and casino business is problematic in the event of worldwide recessions. In 2004, work had begun on the luxury multifunctional City Center complex consisting of several hotels and apartment buildings as well as a shopping mall on a 27 ha site on the Strip ( Fig. 5.49). The intention was to ensure global attention by attracting the world's most renowned architects such as Daniel Libeskind, Norman Foster and César Pelli (Schmid 2009, p. 177). The investors were MGM Resort International and Dubai World. When Dubai World almost had to declare bankruptcy in 2009, it was feared that the project, which was under construction, would have to be abandoned. Dubai World was rescued at the last minute by Abu Dhabi, and the individual buildings opened one by one. The project cost a total of $8.5 billion. The timing was bad, however, as the total of 5900 new hotel rooms were redundant as tourist numbers plummeted at the same time. Condominium sales were also difficult (New York

Fig. 5.49   The new multifunctional City Center

Times 10/17/2012g). A similar fate befell the $3.9 billion Cosmopolitan of Las Vegas, construction of which had begun in 2005, financed by Deutsche Bank. After long fears of a definitive halt to construction, the *resort hotel* with almost 3000 rooms opened at the end of 2010 (Handelsblatt 17/10/2011).

5.11.2 Misery

Since the 1990s, the city's boom seemed unstoppable. More and more fancy fantasy worlds were created and the population virtually exploded. However, in Las Vegas, the proportion of those employed in simple services in the hotels and casinos was always large. Large income disparities exist between those who work on the Strip and those who work at other locations, because on the Strip almost all employees are unionized, while many of the other hotels do not hire union members. In fact, union density is higher on the Strip than in any other city in the U.S., which is why Las Vegas is known as *New Detroit*, because union influence and wages have always been high in the automobile city as well. In the largest *resort hotels*, tens of thousands of workers perform a variety of jobs, most of whom are organized in the Hotel Employees and Restaurant Employees Union or the Culinary Workers Local 226. Modeled on these unions, the Strip's many construction workers have also organized in the Carpenters Union. Union members are predominantly white and earn up to 100% more than unorganized workers in comparable occupations. At the other end of the socioeconomic scale are many of the Hispanic newcomers who feel systematically excluded from the unions and often have to do their work as low-paid day laborers. On the sidewalks of the Strip, they hand out flyers to passersby or help build their

5

own homes. The discrepancy between the luxury and the many rich on the one hand and the day laborers on the other reflects the dark side of the globalized world in Las Vegas in a very confined space (Ventura 2003, pp. 98–110).

For decades, Las Vegas' economy and population grew almost unchecked until the recession of 2008 hit Clark County harder than almost any other region in the United States. The economy was completely overheated and largely monostructured. Gambling revenues, McCarran Airport passenger numbers, and conference attendance dropped dramatically. When the crisis erupted, several hotels with thousands of rooms were under construction. Some of those hotels have not opened to this day, as investors have filed for bankruptcy. Nevertheless, the number of guest rooms has continued to increase (◘ Table 5.7). Since at the same time the number of tourists has collapsed, the occupancy rate has also developed negatively, but was still high at a good 80%. However, this high value could only be achieved by means of low room rates. At 39.7 million, the number of tourists in 2012 was even slightly higher than in 2007, but the proportion of conference visitors, who are particularly sought-after, has fallen in the long term. Large corporations, unions and government agencies used to enjoy holding their conventions in the luxurious hotels on the Strip, as there is no other location in the U.S. where business and pleasure can be better combined. American convention tourism suffered a major setback when it was revealed that insurance conglomerate AIG had held a $440,000 employee conference at the St. Regis Resort Monarch Beach in Dana Point, California, in September 2008. A few days

earlier, AIG had received $85 million from the government to avert bankruptcy. President Obama strongly condemned the trip at taxpayer expense. As a result, conventions planned at luxury hotels nationwide were canceled. In Las Vegas, which was particularly suspected of having been chosen as a location because of its vast leisure offerings, 26% of all scheduled conventions were cancelled in the following 6 months alone, causing $166 million in economic damage. Some of the conventions were even relocated to uninviting Detroit. The negative impact of the luxury binge on convention tourism in particularly attractive locations is referred to as the AIG effect (Las Vegas Review Journal 9/27/2009). With corporations and other organizations still preferring unfamiliar and less expensive locations for their conventions, convention tourism in Las Vegas has not been able to recover to this day.

In 2012, almost half of all jobs in Las Vegas depended on tourism. Most employees are poorly qualified and earn little. The sector is also highly dependent on fluctuations in the economy. Las Vegas has long sought to diversify its economy, but success has been slow in coming. In 2004, online retailer Zappos had moved its headquarters from California to Nevada, and in 2008, the SWITCH Super-Nap server, now the size of 11 football fields, had opened near the airport. State agencies and companies such as eBay and Google manage their data here. Although Nevada's business tax is low, there is no income tax, and the legal requirements are few, the state has not yet succeeded in attracting other major companies. After Washington, D.C., where government jobs are concentrated, and Alaska, which depends on oil and gas, no other state's economy is as

◘ Table 5.7 Selected indicators of the economic development of the Las Vegas MA

Indicator	1990	2007	2010	Change 2007 until 2010 in %	2012
Rooms in hotels and motels	73,730	132,947	148,935	+12	150,481
Utilization rate (in %)	89.1	90.4	80.4	−10	84.4
Tourists (in millions)	20.9	39.2	37.3	−4.8	39.7
Passengers McCarran Airport (in millions)	19.0	47.7	39.8	−16.6	41.7
Gaming revenues (in billions of US dollars)	4.1	10.9	8.9	−18.3	9.4
Visitors to conferences (in million)	1.7	6.2	4.5	−27.4	4.9
House prices in US dollars (median)	k. A.	219,724 (2008)	132,294	−37	139,800

Sources: City of Las Vegas (2012); Zillow Real Estate (2012); ▶ www.fortune.com

undiversified as Nevada's. Las Vegas hopes to expand logistics and medical services for seniors, who have so far preferred to seek care in California. Further stimulus could come from geothermal expansion and unmanned test flights, which the Air Force is already conducting in northern Nevada. Potential investors, however, are few and far between. Clark County's schools and universities don't have a good reputation, and knowledge-intensive services are largely absent. The creative class is not attracted to Las Vegas, and an innovative milieu is lacking (Las Vegas Review Journal 9/21/2011b; The Economist 7/1/2012a).

In the USA, house prices, by which is meant the prices of all housing units, are regarded as an important indicator of the economic development of a region. Parallel to the rapid increase in population, a building boom had begun in the 1980s. New residents initially bought a relatively inexpensive house to sell at a profit once incomes and house prices had risen. The profit, which was usually around 20% of the purchase price, was invested in a better house. Building permits were issued in large numbers with no regard for the environment. In 1998 alone, 32,000 building permits were issued in Las Vegas (Depken et al. 2009, pp. 248–254; Ventura 2003, p. 103). After minor losses beginning in 2005, demand for housing really collapsed in 2008, and home prices fell 37% by 2010. To date, home prices in Las Vegas have barely recovered. At only around $140,000, they are far below other major cities in the western U.S. Analogous to the drop in prices, Nevada held the sad record of having the highest number of foreclosures per 1000 housing units from 2008 to 2012. More than 70% of Nevada's population lives in Las Vegas, which was particularly hard hit by foreclosures. After 62 months, the neighboring state of Arizona recorded minimally more foreclosures than Nevada for the first time in the spring of 2012. However, at that point in Las Vegas, the financial situation of homeowners was still dramatic, with the mortgage amount exceeding the value of homes for 71% of all indebted homeowners. That was twice the national average and also significantly more than in Arizona. For a quarter of all indebted homeowners, the mortgage was even twice the value of the homes. Over-indebted homes and apartments can be sold at great losses at best, if they can find a buyer at all (▶ www.zillow.com). Large segments of the middle class in Las Vegas have suffered large losses of income or even lost their homes in recent years. Homelessness, however, is not officially provided for in Las Vegas. The city wants to present itself to tourists as clean and spotless as possible. Homeless men in particular lack emergency shelters, and they are often punished by the police for very minor offenses such as passing outside marked crosswalks (Borchard 2005). In 2009, more than 13,000 homeless people were counted in Las Vegas. Emergency shelters, however, existed for only a portion of these people, many of whom tried to eke out a living as day laborers (Las Vegas Review Journal 1/27/2011a). On the Strip, where begging could provide a livelihood, even the sidewalks are owned by adjacent property owners. Homeless people are promptly evicted. Many find a place to stay only in the sprawling canal system beneath the city. At the height of the crisis, some 300 people were reportedly living here (Huffington Post 1/4/2010). Even if it is invisible, real misery in Las Vegas is often only a few meters away from the luxurious hyperreal worlds.

The Future of the American City

Many examples were used to illustrate that *"the"* U.S.-American city does not exist. Due to their age, cities on the East Coast differ significantly from those on the West Coast. While the former have a clearly defined center, the latter is no longer guaranteed to have one. Many cities in the Northeast have suffered large population losses in recent decades, but most cities in the West and South have seen large population gains. Despite all the differences, megatrends, each of which applies to a larger number of cities, can be identified:

1. The location of cities has become less important. While the industrial cities of the nineteenth century could only be successful along waterways, today some of the country's largest cities are located in desert regions. Efficient airports that are well integrated into the national and international air transport network are now more important than sea or river ports and even railway stations.

2. Services and creativity are more important than industry for the success of cities.

3. Many cities in the old industrialized Northeast of the US have been losing population for decades, while cities in the West and South have grown, sometimes explosively. Nevertheless, it is too simplistic to contrast shrinking and growing cities, because within these two groups there are major differences in terms of the ethnic composition of the population as well as their schooling and income. This also applies to the creative potential of the cities.

4. Suburban areas continue to grow far faster than core cities.

5. Suburban space is not homogeneous, but very heterogeneous. In many respects, it aligns itself with the core cities in terms of form and function.

6. In the past 50 years, the urban system of the USA has changed fundamentally. Settlements like San Jose, which were almost unknown as recently as the mid-twentieth century, are now among the largest cities in the US, while Detroit and St. Louis are on their way to complete insignificance. At the same time, the scale and nature of the material and immaterial flows between cities have changed fundamentally.

7. Globalization affects all cities. U.S. cities compete globally for the brightest minds and biggest investors. Cities form nodes of transport and communication in a global network that is subject to constant change. A city's future depends on its role in the global network.

8. The vulnerability of cities has increased due to climate change. In addition, terrorist attacks pose a latent threat.

9. Neoliberalism, deregulation and new forms of urban government determine the development of cities.

10. Integrity, assertiveness and vision of mayors determine the success or failure of a city. Cities today are run like large companies, of which the mayors are the managers.

11. Reurbanisation is taking place, which is associated with restructuring.

12. Cities develop according to the laws of the free market or, more precisely, the real estate market.

13. Downtowns are being transformed into showcases for the city and region with flagship stores, expensive museums, luxury hotels, sports stadiums and buildings designed to generate global attention. Festivalization is taking place.

14. Cities have become interchangeable. Many downtowns no longer convey an identity, and shopping centers are similar all over the country anyway. Even the neighborhoods designed on the drawing board all look more or less the same.

15. New urbanites are moving into the expensive apartments and lofts of the inner cities. They use the downtowns differently than their predecessors.

16. The forbidding megastructures of the 1960s–1980s have been replaced by smaller and more open buildings.

17. Since the 1990s, more high-rises have been built in many downtowns than in previous decades. What is new is the large number of high-rise residential buildings.

18. Industrial cities are becoming consumer cities. Cities that do not succeed in this transformation become losers, as they lose the competition for creative minds and global capital.

19. Private capital is essential for urban redevelopment. The implementation of many ideas takes place within the framework of public private partnerships and business improvement districts.

20. There has been a privatization of public space, which is controlled by cameras and private security guards. These also ensure cleanliness and remove undesirables.

21. New neighborhoods are being built by private developers. The proportion of gated communities is large, but the surrounding fences or walls have less influence on daily life than the elaborate rules of the homeowner associations, to which the residents voluntarily submit.

22. Gentrification has become a widespread phenomenon that can be observed in all cities. The gentrification process of dilapidated neighborhoods is publicly promoted. Long-established residents are displaced without new housing being created.

23. The U.S. city is a fragmented city. Different uses lie next to each other seemingly at random.

24. The U.S. city is a segregated city. Affluent and poor Americans live spatially separated from each other.

Whether ethnic segregation is declining is a matter of debate, as Census data leave much room for interpretation.

25. Seniors retire to retirement communities with great recreational opportunities. Low taxes are an important reason for living in retirement communities.

26. Indigent and poor Americans are the losers of neoliberal policies. The dispossessed are marginalized and homelessness is accepted; these groups of the population are ruthlessly marginalized from public spaces.

All of the aforementioned megatrends can be observed simultaneously and often in close proximity to one another in the U.S. city. In recent times, however, a rethink has set in:

1. The city is experiencing a new appreciation in the public and scientific discussion.
2. Public transport is being expanded.
3. The protection of the environment occupies a larger space.
4. Urban sprawl is being combated. Brownfields are being put to new uses. Urban growth bounderies are intended to limit the growth of land, and compact neighborhoods are being created as part of the new urbanism.
5. More attention is being paid to the quality of public space. In the downtowns, small parks are being created in vacant lots or even pedestrian zones.

Whether the new ideas will lead to a paradigm shift and changes in the medium to long term remains to be seen. What is certain, however, is that the U.S. city is not a discontinued model. Cities will continue to be centers of the economy in the coming decades and probably even centuries, and there will always be people who, despite all the disadvantages such as high land prices, noise, and a large ecological burden, prefer the bustling city life to a monotonous country life. Immigrants will also continue to prefer to settle in cities because of the large and diverse range of jobs on offer. It is not yet possible to gauge whether the positive development that many inner cities have experienced in recent years will spill over to the city as a whole. It would certainly be advantageous if core cities and suburbs were no longer hostile to each other but were to cooperate closely in spatial development. Instead of constantly building new roads on the periphery, existing roads should be rehabilitated and public transport should be encouraged. New development should only be permitted within existing residential areas and preferably near public transit stops. Individual communities are always competing to attract amusement parks, shopping centers or office parks, as these facilities promise high tax revenues. Neighboring

communities go away empty-handed, but suffer from the increasing traffic volume. This vicious circle must be broken (Katz and Bradley 1999).

Another question is which city has the greatest potential. Joel Kotkin (2011), together with Mark Schill of Praxis Strategy Group, studied the development prospects of the 50 largest metropolitan areas in the U.S. (◘ Fig. 6.1). They assumed that cities with growing, young, and skilled populations will develop more positively than cities with declining, old, and poorly educated populations. The development of jobs in earlier years was included in the calculations to the extent of about one third. Demographic data such as population growth, the percentage of children between the ages of 5 and 17, and the number of immigrants with at least a college degree were also considered. Austin MA (ranked No. 1) and Raleigh MA (ranked No. 2) fared best, as these regions attract large numbers of immigrants, have young and well-educated populations, and have added a particularly large number of jobs in recent years. Nashville (ranked 3rd) was able to attract many skilled immigrants with a large supply of affordable housing and a good business climate. Washington, DC, MA. (ranked 6th), which spans parts of the bordering states of Virginia, Maryland, and West Virginia, is the northernmost city in the top ten. The U.S. capital has been able to benefit from its proximity to the federal government in recent decades. However, if the administration is indeed going to be scaled back, as Republicans have repeatedly called for, there are fears of a negative impact on the region. Phoenix (ranked 9th) and Orlando (ranked 10th) suffered greatly from the recession triggered in 2008, but still had far more jobs in 2010 than a decade earlier. Los Angeles (ranked 47th), on the other hand, has been particularly negative in more recent times. With manufacturing job losses not being offset, high housing prices, and a poor school system, many families with children migrated out and few young families migrated in. MAs New York (ranked 35th), Chicago (sharing 47th place with Los Angeles) and San Francisco (ranked 42nd) have also disappointed. While these cities will continue to benefit from the financial sector, many corporate headquarters, and good universities, they have rapidly aging populations and are not creating enough new jobs (Kotkin 2011). It is important to note that other studies, with different indicators and weightings, may come to different conclusions. But the trend is clear: the metropolitan areas with good prospects are mostly in the South and West of the US, while the MAs with limited prospects are mostly in the Northeast of the country.

Although the future of the US city cannot be accurately predicted for many reasons, it will certainly remain an interesting field of research.

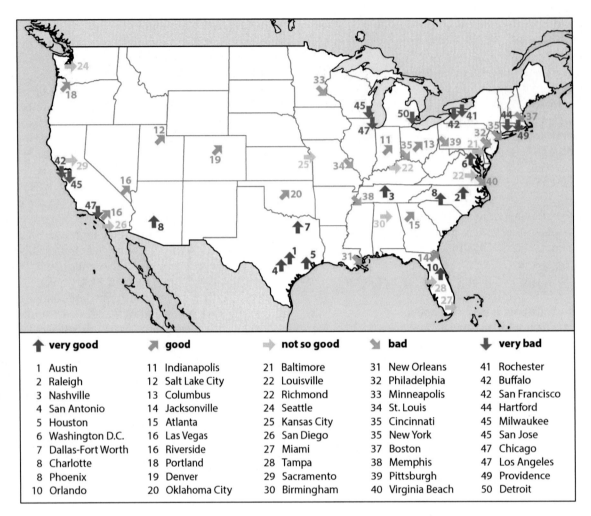

◻ **Fig. 6.1** Development potential of the 50 largest *metropolitan areas.* (Data basis Kotkin 2011)

Supplementary Information

References

Abu-Lughod, Janet L. (1999): New York, Chicago, Los Angeles. America's Global Cites. Minneapolis, London.

Abu-Lughod, Janet L. (2011): The City Revisited: Urban Theory from Chicago, Los Angeles and New York. Minneapolis.

Acker, Kristin (2010): Die US-Expansion des deutschen Discounters Aldi. Eine Fallstudie zur Internationalisierung des Einzelhandels. Geographische Handelsforschung 16. Passau.

Adkins, Julie (2010): The View from the Front Desk. Addressing Homelessness and the Homeless in Dallas. In: Gmelch, George et al. (Ed.): Urban Life: Readings in the Anthropology of the City. Long Grove, Ill., S. 217–231.

Alliance for Downtown New York (2010): Downtown Living in New York City's most Dynamic Neighborhood. New York.

Alliance for Downtown New York (2011): 10 Years later. New York.

Aloisi, James A., Jr. (2004): The Big Dig. Carlisle, Mass.

American Seniors Housing Association (2012): ASHA 50. A Special Supplement to National Real Estate Investors. Cincinnati.

Arens, Robert (2004): The Heidelberg Project. In: Oswalt, Philipp (Ed.): Schrumpfende Städte, Bd. 1. Ostfildern, S. 450–455.

Armstrong, Amy et al. (2007): The Benefits of Business Improvement Districts: Evidence from New York City. Furman Center Policy Brief. New York, July.

Associated Press, Pressemitteilung (2010): Detroit to Bulldoze 40 Square Miles. 8. März.

Associated Press, Pressemitteilung (2012a): Big Dig Pegged at $ 24, 3 B, Lawmakers told. 10. Juli.

Associated Press, Pressemitteilung (2012b): Silicon Valley's Light Rail among least Efficient. 27. Dezember.

Atkinson, Rowland and Sarah Blandy (2006): Introduction: International Perspectives on the New Enclavism and the Rise of Gated Communities. In: Atkinson, Rowland und Sarah Blandy (Ed.): Gated Communities. London, New York, S. i–xvi.

Atlanta Regional Commission (2011): Regional Snapshot. A Decade of Change in the Atlanta Region: A Closer Look at the 2010 Census. Atlanta.

Augé, Marc (2011, 1960): Nicht-Orte. 2. Aufl. München.

Barber, Benjamin R. (2013): If Mayors Ruled the World. Dysfunctional Nations, Rising Cities. New Haven und London.

Baudrillard, Jean (1993): Hyperreal America. In: Economy and Society 22 (2), S. 243–252.

Baudrillard, Jean (1994): Simulacra and Simulation. Ann Arbor.

Baumgarten, Mark (2009): The New Nashville. In: The Next American City 22 (Spring), S. 16–18.

Baum-Snow, Nathaniel (2010): Changes in Transportation and Commuting Patterns in U.S. Metropolitan Areas, 1960–2000. In: American Economic Review 100 (2), S. 378–382.

Beauregard, Robert A. (1986): The Chaos and Complexity of Gentrification. In: Smith, Neil und Peter Williams (Ed.): Gentrification and the City. London, S. 35–55.

Beauregard, Robert A. (2006): When America Became Suburban. Minneapolis.

Beauregard, Robert A. (2011): Radical Uniqueness and the Flight from Urban Theory. In: Judd, R. Dennis und Dick Simpson (Ed.): The City, Revisited. Urban Theory from Chicago, Los Angeles, and New York. Minneapolis, S. 186–202.

Beaverstock, Jonathan et al. (1999): A Roaster of World Cities. In: Cities16 (6), S. 445–458.

Benfield, Kaid (2011): Detroit: The "Shrinking City" that isn't Actually Shrinking. In: The Atlantic 6, Online-Ausgabe, 2 S.

Bennett, Larry (2010): The Third City: Chicago and American Urbanism. Chicago.

Berger, Alan (2007): Drosscape. Wasting Land in Urban America. New York.

Berry, Brian J. L. (1980): Inner City Futures: an American Dilemma Revisited. Transactions of the Institute of American Geographers, New Series 5, S. 1–28.

Bettencourt, Luis M. and Geoffrey B. West (2011): Bigger Cities do more with less. In: Scientific American 9, S. 52–53.

Biello, David (2011): How Green is my City. Retrofitting is the best Way to Clean up Urban Living. In: Scientific American, Sept., S. 66–69.

Binelli, Mark (2011): Don't Shrink Detroit, Super-Size it. In: Atlantic Monthly 3, Online-Ausgabe, 3 S.

Birch, Eugéne (2005): Who Lives Downtown. Brookings Report, Living Cities Census Series. Washington, D.C.

Bishop, Peter and Lesley Williams (2012): The Temporary City. London, New York.

Blakely, Edward and Mary G. Snyder (1999): Fortress America. Gated Communities in the United States. Washington, D.C.

Blechman, Andrew D. (2008): Leisureville: Adventures in America's Retirement Utopias. New York.

Bloomberg, Michael (2011): The Best and the Brightest. New York City's Bid to Attract Science Talent could serve as a Model for Other Cities. In: Scientific American, Sept., S. 16.

Blum, A. (2011): Tunisia, Egypt, Miami: The Importance of Internet Choke Points. In: The Atlantic 1, Online-Ausgabe, 2 S.

Blume, Helmut (1988): USA. Eine Geographische Landeskunde, Bd. II. 2. Aufl., Darmstadt.

Borchard, Kurt (2005): The Word on the Street: Homeless Man in Las Vegas. Las Vegas.

Boswell, T. D. (2000): Cuban Americans. In: McKee, J. O. (Ed.): Ethnicity in Contemporary America. A Geographical Appraisal. Lanham, S. 139–177.

Brake, Klaus (2012): Reurbanisierung – Interdependenzen zum Strukturwandel. In.: Brake, Klaus und Günter Herfert (Ed.): Reurbanisierung. Wiesbaden, S. 22–33.

Brake, Klaus and Günter Herfert (2012): Auf dem Weg zu einer Reurbanisierung? In.: Brake, Klaus and Günter Herfert (Ed.): Reurbanisierung. Wiesbaden, S. 13–19.

Brake, Klaus and Rafael Urbanczyk (2012): Reurbanisierung – Strukturierung einer begrifflichen Vielfalt. In: Brake, Klaus and Günter Herfert (Ed.): Reurbanisierung. Wiesbaden, S. 34–51.

Brenner, Neil and Nik Theodore (2002): Cities and the Geographies of "Actually Existing Neoliberalism". In: Brenner, Neil und Nik Theodore (Ed.): Spaces of Neoliberalism: Urban Restructuring in North America and Western Europe. New York, S. 2–32.

Bridges, Amy (2011): The Sun Also Rises in the West. In: Judd, R. Dennis und Dick Simpson (Ed.): The City, Revisited. Urban Theory from Chicago, Los Angeles, and New York. Minneapolis, S. 79–103.

Brookings Institution Center on Urban and Metropolitan Policy (2004): Income Trends in Miami. Washington, D.C.

Brown, Michael und Richard Morrill (Ed.) (2011): Seattle Geographies. Washington, D.C.

Bruegmann, Robert (2005): Sprawl: A Compact History. Chicago.

Burgess, Ernest W. (1925): The Growth of the City. In: Park et al.: The City: Suggestions for Investigation of Human Behavior in the Urban Environment. Chicago (reprint 1967), S. 47–62.

Burnham, Daniel H. und Edward E. Bennett (1909): Plan of Chicago. Chicago.

Bushwick, Sophie (2011): The Top 10 Cities for Technology. In: Scientific American, 19. Aug., Online-Ausgabe, 3 S.

Button, Kenneth und Roger Stough (2000): Air Transport Networks. Theory and Policy Implications. Cheltenham, UK u. Northampton, MA.

Calbet i Elias et al. (2012): Standortfaktor Innenstadt – Ambivalenzen der Reurbanisierung in Barcelona, London und Chicago. In: Brake, Klaus and Günter Herfert (Ed.): Reurbanisierung. Wiesbaden, S. 388–404.

Campanella, Richard (2007): An Ethnic Geography of New Orleans. In: Journal of American History 94 (12), S. 704–715.

Campanella, Thomas J. (2006): Urban Resilience and the Recovery of New Orleans. In: Journal of the American Planning Association 72 (2), S. 141–146.

Carter, Harold (1980): Einführung in die Stadtgeographie. Berlin, Stuttgart.

Castells, Manuel (2004): Informationalism, Networks, and the Network Society: A Theoretical Blueprint. In: Castells, Manuel (Ed.): The Network Society: A Cross Cultural Perspective. Cheltenham, U.K., S. 3–45.

Caves, Roger W. (Ed.) (2005): Encyclopedia of the City. London u. New York.

Center for Community Progress (2012): Turning Vacant Spaces into Vibrant Places. 2011 Annual Report. Washington, D.C.

Chicago Historical Society (Ed.) (2004): Encyclopedia of Chicago. Chicago.

Chicago Loop Alliance (2011): Loop Economic Study & Impact Report. Chicago.

Chicago Office of Tourism and Culture (2011): 2010 Statistical Information. Chicago.

City of Chicago (2003): The Chicago Central Area Plan. Preparing the Central City for the 21st Century. Draft Final Report to the City of Chicago Plan Commission. Chicago.

City of Las Vegas (2012): Data Book. Las Vegas.

City of New York, Department of City Planning (2002): 2000/2001 Report on Social Indicators. New York.

City of New York, Department of City Planning (2013): Zoning Glossary. New York.

Clark, Terry Nichols (2011): The New Chicago School. Notes towards a Theory. In: Judd, Dennis R. and Dick Simpson (Ed.): The City, Revisited. Urban Theory from Chicago, Los Angeles, and New York. Minneapolis, S. 220–241.

Clark, Terry Nichols (Ed.) (2004): The City as Entertainment Machine: Research in Urban Policy 9. Amsterdam.

Clark, Terry Nichols et al. (2002): Amenities Drive Urban Growth. In: Journal of Urban Affairs 24, S. 493–515.

Clay, P. (1979): Neighborhood Renewal: Middle-Class Resettlement and Incumbent Upgrading in American Neighborhoods. Lexington.

Coates, Ta-Nehesi (2011): The Other Detroit. The City's Grandest Enclave Clings to the Dream. In: The Atlantic 4, S. 20–21.

Cohen, Peter and Fernando Marti (2009): Searching for the "Sweet Spot" in San Francisco. In: Porter, Libby and Kate Shaw: Whose Urban Renaissance? An International Comparison of Urban Regeneration Strategies. London u. New York, S. 222–233.

Colten, Craig E. (2005): An Unnatural Metropolis. Wrestling New Orleans from Nature. Baton Rouge.

Comfort, Louise und Thomas A. Birkland (2010): Retrospectives and Prospectives on Hurricane Katrina: Five Years and Counting. In: Public Administration Review, Sept./Oct., S. 669–678.

Community Association Institute (2013): Greater Los Angeles Chapter Statistics. Los Angeles. (www.cai-glac.org)

Conzen, Michael (1983): Amerikanische Städte im Wandel. Die neue Stadtgeographie der achtziger Jahre. In: Geographische Rundschau 35 (4), S. 142–150.

Conzen, Michael (2010): The Making of the American Landscape. 2. Aufl., New York, London.

Conzen, Michael and Nicholas Dahmann (2006): Re-Inventing Chicago's Core: The Diversity of New Upscale Housing Districts in and near the Loop. AAG 2006 Annual Meeting, Field Trip Guide. Chicago.

Corso, Anthony W. (1983): San Diego. The Anti-City. In: Bernard, Richard M. und Bradley R. Rice (Ed.): Sunbelt Cities: Politics and Growth since World War II. Austin, S. 328–344.

Cox, Wendell (2010): The Downtown Seattle Jobs Rush to the Suburbs. In: www.newgeography.com, 20.4.2010, 2 S.

Cox, Wendell (2011a): Suburbanized Core Cities. www.newgeography.com, 26.8.2011, 5 S.

Cox, Wendell (2011b): The Evolving Urban Form: Chicago. www.newgeography.com, 18.7.2011, 7 S.

Cox, Wendell (2011c): The Accelerating Suburbanization of New York. www.newgeography.com, 29.3.2011, 6 S.

Cox, Wendell (2011d): The Evolving Urban Area: Seattle. www.newgeography.com, 30.6.2011, 4 S.

Crawford, Margaret (1992): The World in a Shopping Mall. In: Sorkin, Michael (Ed.): Variations of a Theme Park. The New American City and the End of Public Space. New York, S. 181–204.

Cremin, Dennis H. (1998): Chicago's Front Yard. In: Chicago History (Spring), Chicago, S. 22–43.

Croucher, Sheila L. (2002): Miami in the 1990s: "City of the Future" or "City on the Edge"? In: Journal of International Migration and Integration 3 (2), S. 223–239.

CTBUH (Council on Tall Buildings and Urban Habitat) (2008): Tall Buildings in Numbers. The Tallest Buildings in the World: Past, Present & Future. In: CTBUH Journal 2, S. 40–41.

CTBUH (Council on Tall Buildings and Urban Habitat) (2011): Tall Buildings in Numbers. New York Skyscrapers. In: CTBUH Journal 3, S. 54–55.

CTBUH (Council on Tall Buildings and Urban Habitat) (2013): Tall Buildings in Numbers. In: CTBUH Journal 1, S. 46–47.

Cullingworth, John B. and Roger W. Caves (2008): Planning in the USA: Policies, Issues, and Processes. 3. Aufl., London u. a.

Curtis, Wayne (2009): Houses of the Future. In: The Atlantic 11, Online-Ausgabe, 10 S.

Davidson, Mark and Loretta Lees (2010): New-Build Gentrification: Its Histories, Trajectories, and Critical Geographies. In: Population, Space and Place 16 (5), S. 395–411.

Davis, Mike (1992a): City of Quartz: Excavating the Future in Los Angeles. New York.

Davis, Mike (1992b): Fortress Los Angeles: The Militarization of Urban Space. In: Sorkin, Michael: Variations of a Theme Park: The New American City and the End of Public Space. New York, S. 154–180.

Davis, Mike (2001): The Flames of New York. In: New Left Review 12, Online-Ausgabe, 9 S.

Davis, Mike (2002): Dead Cities and other Tales. New York.

Dear, Michael J. (2005): Die Los Angeles School of Urbanism. In: Geographische Rundschau 57 (1), S. 30–36.

Dear, Michael J. (Ed.) (2002): From Chicago to L. A.: Making Sense of Urban Theory. Thousand Oaks.

Dear, Michael J. und Nicholas Dahmann (2011): Urban Politics and the Los Angeles School of Urbanism. In: Judd, R. Dennis and Dick Simpson (Ed.): The City, Revisited. Urban Theory from Chicago, Los Angeles, and New York. Minneapolis, S. 65–78.

Dear, Michael J. und Steven Flusty (1998): Postmodern Urbanism. In: Annals of the Association of American Geographers 88 (1), S. 50–72.

Depken, Craig A. et al. (2009): An Empirical Analysis of Residential Property Flipping. In: Journal of Real Estate Finance and Economics 37 (3), S. 265–279.

Deskins, Donald R. (1996): Economic Restructuring, Job Opportunities and Black Social Dislocation in Detroit. In: O'Loughlin, John und Jürgen Friedrichs (Ed.): Post Industrial Cities. Berlin, New York, S. 259–284.

Doel, Marcus A. (1999): Occult Hollywood: Unfolding the Americanization of World Cinema. In: Slater, David und Peter Taylor (Ed.): The American Century. Oxford, S. 243–260.

Dunstan, Roger (1997): Gambling in California. Sacramento.

Erie, Steven P. und Scott MacKenzie (2011): From the Chicago to the L. A. School. In: Judd, R. Dennis und Dick Simpson (Ed.): The City, Revisited. Urban Theory from Chicago, Los Angeles, and New York. Minneapolis, S. 104–134.

Fainstein, Susan S. (2011): Ups and downs of Global Cities. In: Hahn, Barbara and Meike Zwingenberger (Ed.): Global Cities – Metropolitan Cultures. Publikationen der Bayerischen Amerika-Akademie 11, Heidelberg, S. 11–23.

Fainstein, Susan S. and David Gladstone (1999): Evaluating Urban Tourism. In: Judd, Dennis R. and Susan S. Fainstein (Ed.): The Tourist City. New Haven u. London, S. 21–34.

Fainstein, Susan S. and Denis R. Judd (1999): Global Forces, Local Strategies, and Tourism. In: Judd, Dennis R. and Susan S. Fainstein (Ed.): The Tourist City. New Haven u. London, S. 1–17.

Fishman, Robert (1987): Bourgeois Utopias: The Rise and Fall of Suburbia. New York.

Fitch, Robert (1993): The Assassination of New York. New York.

Flint, Anthony (2012): At the 20th Congress for the New Urbanism, a Movement feels its Age. In: The Atlantic 5, Online-Ausgabe, 3 S.

Florida, Richard (2002): The Rise of the Creative Class. London.

Florida, Richard (2005): Cities and the Creative Class. London.

Fogelson, Robert (1967): The Fragmented Metropolis: Los Angeles, 1850–1930. Berkeley.

Forbes Magazine (2012): Quicken Billionaire Dan Gilbert on Giving Back to Detroit. 18. September.

Ford, Kristina (2010): The Trouble with City Planning. What New Orleans Can Teach us. New Haven u. London.

Frank, Robert (2007): Richistan. A Journey though the American Wealth Boom and the Lives of the New Rich. New York.

Frantz, Klaus (2001): Gated Communities in Metro-Phoenix (Arizona). In: Geographische Rundschau 53 (1), S. 12–18.

Freund, David M. P. (2007): Colored Property: State Policy and White Racial Politics in Suburban America. Chicago.

Frieden, Bernard J. and Lynne B. Sagalyn (1989): Downtown Inc., How America Rebuilds its Cities. Boston.

Fussell, Elizabeth (2007): Constructing New Orleans, Constructing Race: A Population History of New Orleans. In: Journal of American History 94, S. 846–855.

Gaebe, Wolf (2004): Urbane Räume. Stuttgart.

Gainsborough, J. F. (2008): A Tale of Two Cities: Civic Culture and Public Policy in Miami. In: Journal of Urban Affairs 30 (4), S. 419–435.

Gallagher, John (2004): Suburbanisierung Detroits. In: Oswalt, Philipp (Ed.): Schrumpfende Städte, Bd. 1, Ostfildern, S. 242–248.

Gallagher, John (2010): Reimaging Detroit. Opportunities for Redefining an American City. Detroit.

Gapp, Paul et al. (1981): The American City. An Urban Odyssey to 11 U.S. Cities. New York.

Garreau, Joel (1991): Edge Cities. Life on the New Frontier. New York u. a.

Gay, Cheri Y. (2001): Detroit Then and Now. San Diego.

Gehl, Jan (2006): New City Spaces. Kopenhagen.

Gelinas, Nicole (2010): Big Easy Rising. Five Years after Katrina, New Orleans are Showing how to do Recovery Right. In: City 3, Online-Ausgabe, 8 S.

Gelinas, Nicole (2013): New York's Sandy Scorecard. In: City 4, Online-Ausgabe, 7 S.

Gerend, Jennifer (2012): U.S. and German Approaches to Regulating Retail Development: Urban Planning Tools and Local Policies. Diss., Würzburg.

Gerhard, Ulrike (2003): Washington, D.C. – Weltstadt oder globales Dorf. In: Geographische Rundschau 55 (1) S. 56–63.

Gerhard, Ulrike (2004): Global Cities. Anmerkungen zu einem Forschungsfeld. In: Geographische Rundschau 56 (4), S. 4–10.

Gerhard, Ulrike (2007): Global City Washington, D.C. Eine politische Stadtgeographie. Bielefeld.

Gerhard, Ulrike (2012): Reurbanisierung – städtische Aufwertungsprozesse in der Global City-Perspektive. In: Brake Klaus and Günter Herfert (Ed.): Reurbanisierung. Wiesbaden, S. 54–86.

Gerhard, Ulrike und Ingo H. Warnke (2007): Stadt und Text. Interdisziplinäre Analyse symbolischer Strukturen einer nordamerikanischen Stadt. In: Geographische Rundschau 59 (7/8), S. 36–42.

Girault, C. (2003): Miami, Capital du Bassin Caraïbe? In: Mappemonde 72 (4), S. 29–36.

Glaeser, Edward (2009): Green Cities, Brown Suburbs. In: City Journal 4, Online-Ausgabe, 6 S.

Glaeser, Edward (2011a): Triumph of the City. Basingstoke u. Oxford.

Glaeser, Edward (2011b): Engines of Innovation. In: Scientific American 9, S. 50–51 u. 54–55.

Glaeser, Edward and Jacob Vigdor (2012): The End of the Segregated Century: Racial Separation in America's Neighborhoods, 1890–2010. Manhattan Institute (Ed.): Civic Report 66, New York.

Gober, Patricia (1989): The Urbanization of the Suburbs. In: Urban Geography 10 (4), S. 311–315.

Gong, Hongmanu and Kevin Keenan (2012): The Impact of 9/11 on the Geography of Financial Services. In: The Professional Geographer 64 (3), S. 370–388.

Gotham, Kevin Fox (2000): Urban Space, Restrictive Covenants and the Origins of Racial Residential Segregation in a US City, 1900–1950. In: International Journal of Urban and Regional Research 24 (3), S. 617–633.

Gotham, Kevin Fox (2012): Disaster, Inc.: Privatization and Post-Katrina Rebuilding in New Orleans. In: Perspectives on Politics 10 (3), S. 633–646.

Gottdiener, Mark (2001): The Theming of America. American Dreams, Media Fantasies, and Themed Environments. Boulder.

Gottdiener, Mark et al. (1999): Las Vegas. The Social Production of an All-American City. Malden.

Gratz, Roberta B. and Norman Mintz (1998): Cities Back from the Edge: New Life for Downtown. New York.

Greater New Orleans Community Data Center (2011): Fewer Jobs mean fewer People and more Vacant Land. New Orleans.

Grey, A. L. (1959): Los Angeles: Urban Prototype. In: Land Economics 35 (3), S. 232–242.

Grosfoguel, R. (1995): Global Logics in the Caribbean City System: the Case of Miami. In: Knox, Paul L. und Peter J. Taylor (Ed.): World Cities in a World System. Cambridge, S. 156–170.

Gruen, Viktor (1973): Das Überleben der Städte. Wege aus der Umweltkrise: Zentren als urbane Brennpunkte. Wien, München, Zürich.

Grünsteidel, Irmtraud and Rita Schneider-Sliwa (1999): Community Garden-Bewegung in New York. In: Geographische Rundschau 51 (4), S. 203–209.

Gyourko, Joseph et al. (2006): Superstar Cities. National Bureau of Economic Research Working Paper 12355. Cambridge, MA.

Hahn, Barbara (1992): Winterstädte. Planung für den Winter in kanadischen Großstädten. Beiträge zur Kanadistik 2. Augsburg.

Hahn, Barbara (1993): Stadterneuerung am Ufer des Ontario-Sees in Toronto. In: Die Erde 124, S. 237–252.

Hahn, Barbara (1996a): Poverty and Affordable Housing in New York City. In: Frantz, Klaus (Ed.): Human Geography in North America. New Perspectives and Trends in Research. Innsbrucker Geographische Studien 26. Innsbruck, S. 115–127.

Hahn, Barbara (1996b): Die Privatisierung des Öffentlichen Raumes in nordamerikanischen Städten. In: Berliner Geographische Arbeiten 44, S. 259–269.

Hahn, Barbara (2001): Erlebniseinkauf und Urban Entertainment Centers. In: Geographische Rundschau 53 (1), S. 19–25.

Hahn, Barbara (2002): 50 Jahre Shopping Center in den USA. Evolution und Marktanpassung. Geographische Handelsforschung 7, Passau.

Hahn, Barbara (2003): Armut in New York. In: Geographische Rundschau 55 (10), S. 50–54.

Hahn, Barbara (2004): New York, Chicago, Los Angeles. Global Cities im Wettbewerb. In Geographische Rundschau 56 (4), S. 12–18.

Hahn, Barbara (2005a): Die USA: Vom Land der Puritaner zum Spielerparadies. In: Geographische Rundschau 57 (1), S. 22–29.

Hahn, Barbara (2005b): Die Zerstörung von New Orleans. Mehr als eine Naturkatastrophe. In: Geographische Rundschau 57 (11), S. 60–62.

Halle, David and Andrew A. Beveridge (2011): The Rise and the Decline of the L. A. and New York Schools. In: Judd, R. Dennis and Dick Simpson (Ed.): The City, Revisited. Urban Theory from Chicago, Los Angeles, and New York. Minneapolis, S. 137–168.

Hamilton County Planning and Development (2011): Community Revitalization Resulting from Stormwater Management Strategies. Indianapolis.

Hanchett, Thomas (1996): U.S. Tax Policy and the Shopping-Center Boom of the 1950 s and 1960 s. In: American Historical Review 10, S. 1083–1111.

Handelsblatt (www.handelsblatt.com) (2011): Die Deutsche Bank verzockt sich in Las Vegas. 17. Oktober.

Hankins, Katherine B. (2009): Retail Concentration and Place Identity: Understanding Atlanta's Changing Retail Landscape. In: Sjoquist, David L. (Ed.): Past Trends and Future Prospects of the American City. The Dynamics of Atlanta. Lanham, S. 85–103.

Hanlon, Bernadette (2010): Once the American Dream. Inner-Ring Suburbs of the Metropolitan United States. Philadelphia.

Harden, Blaine (2004): Brain-Drain Städte in den USA. In: Oswalt, Philipp (Ed.): Schrumpfende Städte, Bd. 1, Ostfildern, S. 178–181.

Harris, Chauncy D. und Edward L. Ullman (1945): The Nature of Cities. Annals of the American Academy of Political and Social Sciences 242, S. 7–17.

Hartshorn, Truman A. (2009): Transportation Issues and Opportunities Facing the City of Atlanta. In: Sjoquist, David L. (Ed.): Past Trends and Future Prospects of the American City. The Dynamics of Atlanta. Lanham, S. 133–159.

Hartshorn, Truman A. und Peter O. Muller (1989): Suburban Downtowns and the Transformation of Metropolitan Atlanta's Business Landscape. In: Urban Geography 10 (4), S. 375–395.

Harvey, David (1996): Justice, Nature and the Geography of Difference. Oxford.

Harvey, David (2003): The Right to the City. In: New Left Review 53, S. 23–40.

Heineberg, Heinz (2006): Stadtgeographie. 3. Aufl., Paderborn u. a.

Helbrecht, Ilse (2005): Geographisches Kapital – das Fundament der kreativen Metropolis. In: Kujath, Hans-Joachim (Ed.): Knoten im Netz. Zur neuen Rolle der Metropolregionen in der Dienstleistungswirtschaft und Wissensökonomie. Reihe Stadt- und Regionalwissenschaften des IRS 4. Münster, Hamburg, London, S. 121–157.

Hill, Richard C. und Joe R. Feagin (1987): Detroit and Houston: Two Cities in Global Perspective. In: Smith, Michael P. und Joe R. Feagin (Ed.): The Capitalist City. London, S. 155–177.

Hirsch, Arnold R. (1983): New Orleans. Sunbelt in the Swamp. In: Bernard, Richard M. und Bradley R. Rice: Sunbelt Cities: Politics and Growth since World War II. Austin, S. 100–137.

Hise, Greg (2002): Industry and the Landscape of Social Reform. In: Dear, Michael J. (Ed.): From Chicago to L. A.: Making Sense of Urban Theory. Thousand Oaks, S. 95–130.

Holcomb, Briavel (1999): Marketing Cities for Tourists. In: Judd, Dennis R. and Susan S. Fainstein: The Tourist City. New Haven u. London, S. 54–70.

Holzner, Lutz (1992): Washington, D.C. – Hauptstadt einer Weltmacht und Spiegel einer Nation. In: Geographische Rundschau 44, S. 352–358.

Homberger, Eric (1994): The Historical Atlas of New York City. New York.

Hoyle, BS et al. (1988): Revitalising the Waterfront. International Dimensions of Dockland Redevelopment. London, New York.

Hoyt, Homer (1933): One Hundred Years of Land Values in Chicago: The Relationship of the Growth of Chicago to the Rise of its Land Values, 1830–1933. Chicago.

Hoyt, Homer (1939): The Structure and Growth of Residential Neighborhoods in American Cities. Washington, D.C.

Hudnut III, William H. (2003): Halfway to Everywhere. A Portrait of America's First-Tier Suburbs. Washington, D.C.

Huffington Post (2010): Tunnels beneath Vegas a Refuge for Homeless. 1. April.

Huffington Post (2012): Broderick Tower Renovation: Detroit Landmark Shows Off Swank Apartments. 5. April.

Husock, Howard (2009): Jane Jacobs's Legacy. In: City Journal, Juli, Online-Ausgabe, 4 S.

International Council of Shopping Centers (2012): Data Bank Shopping Centers. New York. (www.icsc.org)

Jackson, Kenneth T. (1985): Crabgrass Frontier. The Suburbanization of the United States. New York.

Jackson, Kenneth T. (1995): The Encyclopedia of New York. Yale.

Jacobs, Jane (1961): The Death and Life of Great American Cities. New York.

Jaret, Charles et al. (2009): Atlanta's Future: Convergence or Divergence with Other Cities. In: Sjoquist, David L. (Ed.): Past Trends and Future Prospects of the American City. The Dynamics of Atlanta. Lanham, S. 13–47.

Jayne, Mark (2006): Cities and Consumption. London. Milton Park.

Judd, Dennis R. (2011): Theorizing the City. In: Judd, R. Dennis und Dick Simpson (Ed.): The City, Revisited. Urban Theory from Chicago, Los Angeles, and New York. Minneapolis, S. 3–20.

Judd, Dennis R. und Dick Simpson (Ed.) (2011): The City, Revisited. Urban Theory from Chicago, Los Angeles, and New York. Minneapolis.

Kasarda, John D. und Greg Lindsay (2011): Aerotropolis. The Way we Live Next. New York.

Katz, Bruce (2009): Seattle's Opportunity Emerging from the Great Recession. In: The Seattle Times, 12. Okt.

Katz, Bruce und David Jackson (2005): The Great City (Seattle). In: The Seattle Times, 30. Jan.

Katz, Bruce und Jennifer Bradley (1999): Divided we Sprawl. In: The Atlantic 12, Online-Ausgabe, 6 S.

Kavage, Sarah (2004): Today's Company Town: Seattle's Boeing Fixation. In: The Next American City 4, S. 36–39 u. 47.

Kayden, Jerold (2000): Privately Owned Public Space: The New York City Experience. New York.

Keil, Roger (2011): Global Cities: Connectivity, Vulnerability, and Resilience. In: Hahn, Barbara and Meike Zwingenberger: Global Cities – Metropolitan Cultures. Publikationen der Bayerischen Amerika-Akademie 11, Heidelberg, S. 41–63.

Kelling, George L. and James Q. Wilson (1982): Broken Windows. The Police and Neighborhood Safety. In: The Atlantic 3, Online-Ausgabe, 11 S.

Kennedy, Lawrence W. (1992): Planning the City upon the Hill. Boston since 1630. Amherst.

Keys, E. et al. (2007): The Spatial Structure of Land Use from 1970–2000 in the Phoenix, Arizona, Metropolitan Area. In: The Professional Geographer 59 (1), S. 131–147.

Kirk, P. Annie (2009): Naturally Occurring Retirement Communities. Thriving through Creative Retrofitting. In: Abbott, Pauline, S. et al. (Hg): Re-Creating Neighborhoods for Successful Aging. Baltimore, S. 115–143.

Klingle, Matthew (2006): The Emerald City. New Haven.

Kneebone, Elizabeth (2013): Job Sprawl Stalls. The Great Recession and Metropolitan Employment Location. Brookings Metropolitan Opportunity Series. Washington, D.C.

Kneebone, Elizabeth und Emily Carr (2010): The Suburbanization of Poverty: Trends in Metropolitan America, 2000 to 2008. Brookings Institution: Metropolitan Policy Program, Washington, D.C.

Knox, Paul L. (2005): Vulgaria: The Re-enchantment of Suburbia. In: Opolis 1 (2), S. 33–46.

Knox, Paul L. and Linda M. McCarthy (2012): Urbanization: An Introduction to Urban Geography. 3. Aufl., Glenview.

Kolb, Carolyn (2006): Crescent City, Post-Apocalypse. In: Technology and Culture 47 (1), S. 108–111.

Konig, Michael (1982): Phoenix in the 1950s, Urban Growth in the Sunbelt. In: Journal of the Southwest 24 (1), S. 19–38.

Kotkin, Joel (2001): Older Suburbs: Crabgrass Slums or New Urban Frontier? Policy Study 285. Los Angeles.

Kotkin, Joel (2011): The Next Boom Towns in the U.S. www. newgeography.com, 6.7.2012

Kotkin, Joel (2012): Seattle is Leading an American Manufacturing Revival-Top Manufacturing Growth Regions. www. newgeography.com, 24.5.2012

Kotkin, Joel and Fred Siegel (2000): Digital Geography: The Remaking of City and Countryside in the New Economy. Indianapolis.

Kotkin, Joel and Fred Siegel (2004): Too Much Froth: The Latte Quotient is a Bad Strategy for Building Middle Class Cities. In: Siegel, Fred and Harry Siegel (Ed.): Urban Society, New York, S. 56–57.

Kotkin, Joel und Wendell Cox (2011): Cities and the Census. In: City Journal 4, Online-Ausgabe, 3 S.

Kowinski, William S. (1985): The Malling of America. New York.

Kromer, John (2010): Fixing Broken Cities. The Implementation of Urban Development Strategies. New York.

Kujath, Hans-Joachim (2005): Die neue Rolle der Metropolregionen in der Wissensökonomie. In: Kujath, Hans-Joachim (Ed.): Knoten im Netz. Zur Neuen Rolle der Metropolregionen in der Dienstleistungswirtschaft und Wissensökonomie. Reihe Stadt- und Regionalwissenschaften des IRS 4, Münster, Hamburg, London, S. 23–63.

Laaser, Claus-Friedrich and Rüdiger Soltwedel (2005): Raumstrukturen der New Economy – Tod der Distanz? Niedergang der Stadt? In: Kujath, Hans-Joachim (Ed.): Knoten im Netz. Zur Neuen Rolle der Metropolregionen in der Dienstleistungswirtschaft und Wissensökonomie. Reihe Stadt- und Regionalwissenschaften des IRS 4, Münster, Hamburg, London, S. 65–107.

Lacey, Robert (1987): Ford. Eine amerikanische Dynastie. Düsseldorf, Wien, New York.

Lai, Clement (2012): The Racial Triangulation of Space: The Case of Urban Renewal in San Francisco's Fillmore District. In: Annals of the Association of American Geographers 102 (1), S. 151–170.

Lamster, Mark (2011): Castles in the Air. In: Scientific American 9, S. 76–83.

Lang, Robert E. (2003): Edgeless Cities: Exploring the Elusive Metropolis. Washington, D.C.

Lang, Robert E. et al. (2009): Beyond Edge City: Office Geography in the New Metropolis. In: Urban Geography 30 (7), S. 726–755.

Larson, George A. und Jay Pridmore (1993): Chicago Architecture and Design. New York.

Las Vegas Review Journal (www.reviewjournal.com) (2009): "AIG effect" sending meeting to Motown. 27. September.

Las Vegas Review Journal (www.reviewjournal.com) (2011a): Count Reveals many Homeless in Las Vegas. 27. Januar.

Las Vegas Review Journal (www.reviewjournal.com) (2011b): Switch Communications data center expanding to 2.2 million sq. feet. 21. September.

Lees, Loretta (2003): Super-Gentrification: The Case of Brooklyn Heights, New York City. In: Urban Geography 40 (12), S. 487–509.

Lees, Loretta (2008): Gentrification and Social Mixing: Towards an Inclusive Urban Renaissance. In: Urban Studies 45 (12), S. 2944–2470.

Lees, Loretta et al. (2008): Gentrification. New York, London.

Leinberger, Christopher B. und Charles Lockwood (1986): How Business is Reshaping America. In: Atlantic Monthly 258 (10), S. 43–52.

Leiper, Neil (1989): Tourism and Gambling. In: GeoJournal 19 (3), S. 269–275.

Leitner, Helga et al. (2007): Contesting Urban Futures. Decentering Neoliberalism. In: Leitner, Helga (Ed.): Contesting Neoliberalism: Urban Frontiers. New York, S. 1–25.

Lewis, Kenneth und Nicholas Holt (2011): One World Trade Center. In: CTBUH Journal 3, S. 14–19.

Li, Wei (2009): Ethnoburb: The New Ethnic Community in Urban America. Honolulu.

Lichtenberger, Elisabeth (1999): Die Privatisierung des öffentlichen Raumes in den USA. In: Weber, Gerlind (Ed.): Raummuster – Planerstoff. Festschrift für Fritz Kastner zum 85. Geburtstag. Institut für Raumplanung und ländliche Neuordnung der BOKU. Wien, S. 29–39.

Lind, Diana (2009): A Battle for Public Space. In: The Next American City 22 (Spring), S. 24–25.

Low, Setha M. (2010): The Edge and the Center. Gated Communities and the Discourse of Urban Fear. In: Gmelch, George et al. (Ed.): Urban Life: Readings in the Anthropology of the City. Long Grove, Ill., S. 131–143.

Lucy, William und David L. Phillips (2000): Confronting Suburban Decline: Strategic Planning for Metropolitan Renewal. Washington, D.C.

Lunday, Elizabeth (2009): Shrinking Cities U.S.A. In: Urban Land, Nov./Dec., S. 68–71.

Lynn, David und Tim Wang (2008): The U.S. Housing Opportunity: Investment Strategies. In: Real Estate Issues 33 (2), S. 33–51.

Lyons, Heather (2011): Responding to Hard Times in the "Big Easy": Meeting the Vocational Needs of Low-Income African American New Orleans Residents. In: The Career Development Quarterly 59 (4), S. 290–301.

MacDonald, Heather (2010): The Sidewalks of San Francisco. In: City Journal 3, Online–Ausgabe, 13 S.

Madden, David (2010): Revisiting the End of Public Space: Assembling the Public in an Urban Park. In: City Community 9 (2), S. 187–207.

Malanga, Steven (2004): The Curse of the Creative Class. In: City Journal 4, Online–Ausgabe, 8 S.

Malanga, Steven (2010): The Next Wave of Urban Reform. In: City Journal 3, Online-Ausgabe, 9 S.

Manzi, Tony and Bill Smith-Bowers (2006): Gated Communities as Club Goods: Segregation or Social Cohesion? In: Atkinson, Rowland and Sarah Blandy (Ed.): Gated Communities. London, New York, S. 153–166.

Marquardt, Nadine and Henning Füller (2008): Die Sicherstellung von Urbanität. Ambivalente Effekte von BIDs auf soziale Kontrolle in Los Angeles. In: Pütz, Robert (Ed.): Business Improvement Districts. Ein neues Governance-Modell aus Perspektive von Praxis und Stadtforschung. Geographische Handelsforschung 14, Passau, S. 119–136.

Mayer, Harold M. and Richard C. Wade (1969): Chicago. Growth of a Metropolis. Chicago.

McDonald, John F. (2007): Rebuilding New Orleans: Editor's Introduction. In: Journal of Real Estate Literature. 15 (2), S. 199–212.

McKelvey, Blake (1968): The Emergence of Metropolitan America 1915–1966. New Brunswick.

McKenzie, Brian (2010): Public Transportation Usage among U.S. Workers: 2008 and 2009. Washington, D.C.

McKenzie, Evan (1994): Privatopia. Homeowner Associations and the Rise of Residential Private Government. New Haven, London.

Meyer, Christiane und Christian Muschwitz (2008): Detroit. The Motor City's Decline and its Revitalization. In: Geographische Rundschau. International Edition 4 (2), S. 30–34.

Miami International Airport (2010): Cargo Hub 2010–2011. Miami.

Miller, Bradford (2002): Digging up Boston: The Big Dig Builds on Centuries of Geological Engineering. In: Geotimes 10, Online-Ausgabe, 6 S.

Miller, Donald F. (1997): City of the Century. The Epic of Chicago and the Making of America. New York.

Mitchell, Don (2003): The Right to the City: Social Justice and the Fight for Public Space. New York.

Moen, Ole O. (2004): Mobilitätsdrang in Amerika. In: Oswalt, Philipp (Ed.): Schrumpfende Städte, Bd. 1, Ostfildern, S. 198–205.

Mohl, Raymond (1983): Miami. The Ethnic Couldron. In: Bernard, Richard M. u. Bradley R. Rice (Ed.): Sunbelt Cities: Politics and Growth Since World War II. Austin, S. 58–99.

Mollenkopf, John Hull (2011): School is Out. The Case of New York City. In: Judd, R. Dennis and Dick Simpson (Ed.): The City, Revisited. Urban Theory from Chicago, Los Angeles, and New York. Minneapolis, S. 169–185.

Moses, Robert (1962): Are Cities Dead? In: The Atlantic 1, Online-Ausgabe, 5 S.

Muller, Edward K. (2010): Building American Cityspaces. In: Conzen, Michael (Ed.): The Making of the American Landscape, New York u. London, 2. Aufl., S. 303–328.

Myers, Dowell (2002): Demographic Dynamism in Los Angeles, Chicago, New York, and Washington, D.C. In: Dear, Michael J. (Ed.): From Chicago to L. A.: Making Sense of Urban Theory. Thousand Oaks, S. 17–53.

Newman, Harvey K. (1999): Southern Hospitality: Tourism and the Growth of Atlanta. Tuscaloosa.

Newman, Peter und Andy Thornley (2005): Planning World Cities. Globalization and Urban Politics. London.

New York City, Department of City Planning (2013): Zoning Glossary. New York. (www.nyc.gov)

New York Times (NYTimes) (www.nytimes.com) (1981): Obituary. Robert Moses, Master Builder, is dead at 92. 30. Juli.

New York Times (NYTimes) (www.nytimes.com) (2005): Map: Flooding of Hurricene Katrina, 9. September.

New York Times (NYTimes) (www.nytimes.com) (2006): New Orleans, Ex-Tenants Fight for Projects. 26. Dezember.

New York Times (NYTimes) (www.nytimes.com) (2010): Car Ownership by the Hour. 10. September.

New York Times (NYTimes) (www.nytimes.com) (2011): This Lady Liberty is a Las Vegas Teenager. 14. April.

New York Times (NYTimes) (www.nytimes.com) (2012a): Bonuses on Wall St. Expected to Edge up. 4. November.

New York Times (NYTimes) (www.nytimes.com) (2012b): The Gated Community Mentality. 29. März.

New York Times (NYTimes) (www.nytimes.com) (2012c): Rate of Killings Rises 38 Percent in Chicago in 2012. 25. Juni.

New York Times (NYTimes) (www.nytimes.com) (2012d): Must Haves for the Micro-Pad. 11. Juli.

New York Times (NYTimes) (www.nytimes.com) (2012e): Shrink to Fit. Living Large in Tiny Spaces. 21. September.

New York Times (NYTimes) (www.nytimes.com) (2012g): Las Vegas Project Survives a Case of Bad Timing. 17. October.

New York Times (NYTimes) (www.nytimes.com) (2013): De Blasio is Elected New York City Mayor in Landslide. 5. November.

Nijman, Jan (1997): Globalization to a Latin Beat: The Miami Growth Machine. In: The Annals of the American Academy 551, S. 164–177.

Nijman, Jan (2000): The Paradigmatic City. In: Annals of the Association of American Geographers, 90 (1), S. 135–145.

Nijman, Jan (2007): Place-Particularity and "Deep Analogies": A Comparative Essay on Miami's Rise as a World City. In: Urban Geography 28 (1), S. 92–107.

Onboard Media (2002): New York-New York Hotel and Casino: The Greatest City in Las Vegas. Las Vegas.

Osman, Suleiman (2011): The Invention of Brownstone Brooklyn: Gentrification and the Search for Authenticity in Postwar New York. Oxford.

Ott, Thomas (2001): Cascadia: Zukunftswerkstatt oder verlorenes Paradies? In: Geographische Rundschau 53 (19), S. 4–11.

Owen, David (2009): Green Metropolis: Why Living Smaller, Living Closer, and Driving Less are the Keys to Sustainability. New York.

Palmer Woods Association (2011): History of Palmer Woods. (www.palmerwoods.org)

Park, Robert E. et al. (1925): The City: Suggestions for Investigation of Human Behavior in the Urban Environment. Chicago (reprint 1967).

Peck, Jamie and Adam Tickell (2007): Conceptionalizing Neoliberalism, Thinking Thatcherism. In: Leitner, Helga et al. (Ed.): Contesting Neoliberalism: Urban Frontiers. New York, S. 26–50.

Perez, L. (1986): Cubans in the United States. In: The Annals of the American Academy of Political and Social Sciences 487, S. 126–137.

Piiparinen, Richey (2013): The Psychology of the Creative Class. www.newgeography.com, 12.3.2013.

Plensa, Jaume (2008): The Crown Fountain. Ostfildern.

Plunz, Richard (1996): Detroit is Everywhere. In: Architecture Magazine 85, Heft 4, S. 55–61.

Plyer, Allison (2011): Population Loss and Vacant Housing in New Orleans Neighborhoods. Greater Community Data Center. New Orleans.

Port of Miami (2010a): Port Statistics. Miami.

Port of Miami (2010b): 2010–2011 Cruise Guide. Miami.

Portes, Alejoandro und Alex Stepick III (1993): City on the Edge: The Transformation of Miami. Berkeley.

Postrel, Virginia (2007a): Lofty Ambitions. In: The Atlantic 4, Online-Ausgabe, 3 S.

Postrel, Virginia (2007b): A Tale of Two Homes. In: The Atlantic 11, Online-Ausgabe, 4 S.

Pouder, Richard W. And J. Dana Clark (2009): Formulating Strategic Direction for a Gated Residential Community. In: Property Management 27 (4), S. 216–227.

Pries, Martin (2001): Wiederentdeckung der Downtown New York/ Lower Manhattan. In: Geographische Rundschau 53 (1), S. 26–32.

Pries, Martin (2002): New York und die Ereignisse des 11. September 2001. In: Geographische Rundschau 54 (11), S. 61–64.

Pries, Martin (2005): New York City und der 11. September – drei Jahre danach. In: Geographische Rundschau 57 (1), S. 4–12.

Pries, Martin (2009): Waterfront im Wandel: Baltimore und New York. Mitteilungen der Geographischen Gesellschaft Hamburg 100. Hamburg.

Pütz, Robert (2008): Business Improvement Districts als Modell subkommunaler Governance: Internationalisierungsprozesse und Forschungsfragen. In: Pütz, Robert (Ed.): Business Improvement Districts. Ein neues Governance-Modell aus Perspektive von Praxis und Stadtforschung. Geographische Handelsforschung 14, Passau, S. 7–20.

Quigley, Bill und Davida Finger (2012): Katrina Pain Index 2012: Seven Years and Counting. In: The Louisiana Weekly, 27. August, S. 1 u. 9.

Raeithel, Gert (1995): Geschichte der nordamerikanischen Kultur. 3 Bde., Frankfurt am Main.

Rand McNally (1998): Chicago and Vicinity. Chicago.

Ratti, Carlo And Anthony Townsend (2011): Smarter Cities, The Social Nexus. In: Scientific American 9, S. 42–48.

Reardon, Sean F. And Kendra Bischoff (2010): Income Inequality and Income Segregation. Paper presented at the University of Chicago's Sociology. Workshop.

Renn, Aaron (2009): Detroit: Urban Laboratory and the new American Frontier. www.new.geography.com, 11.4.2009.

Rice, Bradley (1983): Atlanta. If Dixie were Atlanta. In: Bernard, Richard M. und Bradley R. Rice (Ed.): Sunbelt Cities: Politics and Growth Since World War II. Austin, S. 31–57.

Ritzer, George (2003): Islands of the Living Dead: The Social Geography of McDonaldization. In: American Behavioral Scientist 47 (2), S. 119–136.

Ritzer, George und Todd Stillman (2004): The Modern Las Vegas Casino-Hotel: The Paradigmatic New Means of Consumption. In: M@n@gement 4 (3), S. 83–99.

Roman, James (2010): Chronicles of Old New York. Exploring Manhattan's Landmark Neighborhoods. New York.

Ross, Bob und Don Mitchell (2004): Neoliberal Landscapes of Deception. Detroit, Ford Field, and The Ford Motor Company. In: Urban Geography 25 (7), S. 685–690.

Ross, Glenwood et al. (2009): Tracking the Economy of the City of Atlanta: Past Trends and Future Prospects. In: Sjoquist, David L. (Ed.): Past Trends and Future Prospects of the American City. The Dynamics of Atlanta. Lanham, S. 51–83.

Rothe, Eugenio M. und Andrés J. Pumariega (2008): The New Face of Cubans in the United States: Cultural Process and Generational Change in an Exile Community. In: Journal of Immigrant & Refugees Studies, Vol. 6 (2), S. 247–266.

Rothman, Hal (2002): Neon Metropolis. New York.

Rothman, Hal und Mike Davis (2002): The Grit beneath the Glitter. Berkeley u. Los Angeles.

Rothstein, Richard (2012): Racial Segregation Continues, and even Intensifies: Manhattan Institute Report Heralding the "End" of Segregation Uses a Measure that Masks Important Demographic and Economic Trends. University of California, Economic Policy Institute, Berkeley.

Rudel, Thomas K. et al. (2009): From Middle to Upper Class Sprawl? Land Use Control and Changing Patterns of Real Estate Development in Northern New Jersey. In: Annals of the Association of American Geographers 101 (3), S. 609–624.

Ruesga, Candida (2000): The Great Wall of Phoenix? Urban Growth Boundaries and Arizona's Affordable Housing Market. In: Arizona State Law Journal 32, S. 1063–1085.

Rushton, Michael (2009): The Creative Class and Economic Growth in Atlanta. In: Sjoquist, David L. (Ed.): Past Trends and Future Prospects of the American City. The Dynamics of Atlanta. Lanham, S. 163–180.

Rybczynski, Witold (1996): City Life. New York.

Sassen, Saskia (1993): Global City: Internationale Verflechtungen und ihre innerstädtischen Effekte. In: Häußermann, Hartmut und Werner Siebel (Ed.): New York. Strukturen einer Metropole. Frankfurt am Main, S. 71–90.

Sassen, Saskia (1994): Cities in a World Economy. Thousand Oaks.

Sassen, Saskia (2002): Global Networks. Linked Cities. New York.

Sassen, Saskia And Frank Roost (1999): The City. Strategic Site for the Global Entertainment Industry. In: Judd, Dennis R. And Susan S. Fainstein (Ed.): The Tourist City. New Haven u. London, S. 54–70.

Schemionek, Christoph (2005): New Urbanism in US-amerikanischen Städten. Ein effektives Planungskonzept gegen Urban Sprawl. Diss., Würzburg.

Schmid, Heiko (2009): Economy of Fascination. Dubai and Las Vegas as Themed Urban Landscapes. Urbanization of the Earth 11, Berlin u. Stuttgart.

Schmidt, Suntje (2005): Metropolregionen als Hubs globaler Kommunikation und Mobilität in einer wissensbasierten Wirtschaft? In: Kujath, Hans-Joachim (Ed.): Knoten im Netz. Zur neuen Rolle der Metropolregionen in der Dienstleistungswirtschaft und Wissensökonomie. Reihe Stadt- und Regionalwissenschaften des IRS 4, Münster, Hamburg, London, S. 285–320.

Schneider-Sliwa, Rita (1999): Nordamerikanische Innenstädte der Gegenwart. In: Geographische Rundschau 51 (1), S. 44–51.

Schneider-Sliwa, Rita (2005): USA. Geographie, Geschichte, Wirtschaft, Politik. Darmstadt.

Schoenfeld, Bruce (2009): The Empty Arena. If you Built it, they might not Come. In: The Atlantic 5, Online-Ausgabe, 2 S.

Schulz, Max (2008): California's Potemkin Environmentalism. In: City Journal 2, Online-Ausgabe, 7 S.

Schwarz, Benjamin (2010). Gentrification and its Discontents. In: The Atlantic 6, Online-Ausgabe, 5 S.

Scott, Allan J. (2002): Industrial Urbanism in Late-Twentieth-Century Southern California. In: Dear, Michael J. (Ed.): From Chicago to L. A.: Making Sense of Urban Theory. Thousand Oaks, S. 163–179.

Short, John R. (2007): Liquid City. Megalopolis and the Contemporary Northeast. Washington, D.C.

Simpson, Dick und Tom Kelly (2011): The New Chicago School of Urbanism and the New Daley Machine. In: Judd, R. Dennis und Dick Simpson (Ed.): The City, Revisited. Urban Theory from Chicago, Los Angeles, and New York. Minneapolis, S. 205–219.

Sinclair, J. (2003): "The Hollywood of Latin America": Miami as Regional Center in Television Trade. In: Television & New Media 4 (3), S. 211–229.

Sjoquist, David L. (2009): Past Trends and Future Prospects of the American City. The Dynamics of Atlanta. Lanham.

Smith Travel Research (2012): Greater Miami and the Beaches 2011 Visitor Industry Overview. Miami.

Smith, Neil (1982): Gentrification and Uneven Development. In: Economic Geography 58 (2), S. 139–155.

Smith, Neil (1986): Gentrification, the Frontier, and the Restructuring of Urban Space. In: Smith, Neil And Peter Williams (Ed.): Gentrification and the City. London, S. 15–34

Smith, Neil (1987): Gentrification and the Rent Gap. In: Annals of the Association of American Geographers 77 (3), S. 462–465.

Smith, Neil (2006): Gentrification Generalized: From Local Anomaly to Urban "Regeneration" as a Global Urban Strategy. In: Fisher, Melissa And Greg Downey (Ed.): Frontiers of Urban Capital. Ethnographic Reflections on the New Economy. Duke, S. 191–208.

Smith, Neil (2010): After Tomkins Square Park. Degentrification and the Revanchist City. In: Bridge, Gary and Sophie Watson (Ed.): The Blackwell City Reader. 2. Aufl., Malden MA u. Oxford, UK, S. 201–210.

Smith, Neil and Peter Williams (1986): Alternatives to Orthodoxy: Invitation to a Debate. In: Smith, Neil and Peter Williams (Ed.): Gentrification of the City. New York, 3. Aufl. 2007.

Sohmer, Rebecca R. and Robert E. Lang (2003): Downtown Rebound. In: Katz, Bruce and Robert E. Lang (Ed.): Redifining Urban and Suburban America. Evidence from Census 2000. Washington, D.C., S. 63–74.

Soja, Edward W. (1992): Inside Exopolis: Scenes from Orange County. In: Sorkin, Michael: Variations of a Theme Park: The New American City and the End of Public Space. New York, S. 94–122.

Soja, Edward W. (1996): Thirdspace: Journeys to Los Angeles and Other Real and Imagined Places. Oxford.

Sokolovsky, Jay (2010). Civic Ecology, Urban Elders, and the New York City's Community Garden Movement. In: Gmelch, George et al. (Ed.): Urban Life: Readings in the Anthropology of the City. Long Grove, Ill., S. 243–255.

Sorkin, Michael (1992): Variations of a Theme Park: The New American City and the End of Public Space. New York.

Sorkin, Michael (2009): Twenty Minutes in Manhattan. New York.

Souther, Mark J. (2007): The Disneyfication of New Orleans: The French Quarter as Façade in a Divided City. In: Journal of American History 94, S. 804–811.

Spirou, Costas (2006): Urban Beautification: The Construction of a New Identity of Chicago. In: Koval, John et al. (Ed.): The New Chicago. A Social and Cultural Analysis. Philadelphia, S. 295–302.

Spirou, Costas (2011): Both Center and Periphery. Chicago's Metropolitan Expansion and the New Downtown. In: Judd, Dennis R. and Dick Simpson (Ed.): The City, Revisited. Urban Theory from Chicago, Los Angeles, and New York. Minneapolis, S. 273–305.

Spivak, Jeffrey (2008): Edgeless Cities. In: Urban Land 4, S. 131–132.

Steffens, Lincoln (1904): The Shame of Cities. New York.

Sugrue, Thomas (1996): The Origins of the Urban Crisis. Race and Inequality in Postwar Detroit. 2. Aufl., Princeton.

Sugrue, Thomas J. (2004): Niedergang durch Rassismus. In: Oswalt, Philipp (Ed.): Schrumpfende Städte. Bd. 1, S. 230–237.

Sun Cities Area Historical Society (2010): Reshaping Retirement in America. Sun City, Arizona. Sun City.

Texas Transportation Institute (2011): Annual Urban Mobility Report. College Station.

The American Assembly (2011): Reinventing America's Legacy Cities: Strategies for Cities Losing Population. Detroit.

The City Club of Portland (1999): Increasing Density in Portland. Portland.

The Economist (2011): The Parable of Detroit. So cheap, there's hope. 22. Oktober.

The Economist (2012a): Diversifying Las Vegas. Rolling the Dice. 7. Januar.

The Economist (2012b): The last Kodak Moment? 14. Januar.

The Economist (2013): Saving Detroit. Iron Orr. 22. Juni.

Thoreau, Henry David (1854): Walden, or Life in the Woods. New York (reprint 1969).

Times Square Alliance (2012): 1992 Times Square Alliance 2012. New York.

Transition New Orleans Task Force (2010): Economic Development. New Orleans.

Ture, Norman B. (1967): Accelerated Depreciation in the United States 1954–60. Washington, D.C.

U.S. Census Bureau (1960): Statistical Abstract of the United States. Washington, D.C.

U.S. Census Bureau (1991): Statistical Abstract of the United States. Washington, D.C.

U.S. Census Bureau (2011a): Urban Area Criteria for the 2010 Census. Federal Register 76 (164). Washington, D.C.

U.S. Census Bureau (2011b): The 2012 Statistical Abstract. The National Data Book. Washington, D.C.

U.S. Census Bureau (2012): Patterns of Metropolitan and Micropolitan Population Change: 2000 to 2010. 2010 Census Special Report. Washington, D.C.

U.S. Department of Justice (Ed.) (2009): The Private Security Industry: A Review of the Definitions, Available Data Sources, and Paths Moving Forward. Washington, D.C.

Vanderkam, Laura (2011): Parks and Recreation. How Private Citizens Saved New York's Public Spaces. In: City Journal 2, Online-Ausgabe, 5 S.

Ventura, Patricia (2003): Learning from Globalization-Era Las Vegas. In: Southern Quarterly 42 (1), S. 97–111.

Vigdor, Jacob (2008): The Economic Aftermath of Hurricane Katrina. In: The Journal of Economic Perspectives 22 (4), S. 135–154.

Wacquant, Loic (2008): Urban Outcasts. A Comparative Sociology of Advanced Marginality. Cambridge, UK, u. Malden, USA.

Whyte, William H. (1988): Rediscovering the Center. New York.

Wilson, William Julius (1987): The Inner City, the Underclass, and Public Policy. Chicago.

Winters, Marcus A. (2010): The Life-Changing Lottery. In: City Journal 3, Online-Ausgabe, 5 S.

Wirth, Louis (1925): A Bibliography of the Urban Community. In: Park et al. (Ed.): The City: Suggestions for Investigation of Human Behavior in the Urban Environment. Chicago (reprint 1967), S. 161–228.

Zolkos, Rodd (2012): New Orleans Boosts Flood Defenses. In: Business Insurance 46 (2), Oline-Ausgabe, 4 S.

Zorbaugh, Harvey W. (1929): The Gold Coast and the Slum: A Sociological Study of Chicago's New North Side. Chicago.

Zukin, Sharon (1982): Loft Living: Culture and Capital in Urban in Urban Change. New Brunswick, N.J.

Zukin, Sharon (2010). Naked City: The Death and Life of Authentic Urban Places. Oxford.

Internet Resources

Alliance for Downtown New York: www.downtowny.com
American Association of Port Authorities: www.aapa-ports.org
American Recovery and Reinvestment Act: www.recovery.gov
Ave Maria, Fl.: www.avemaria.com
Ayn Rand Institute: www.principlesofafreesociety.com
Baltimore Development Corporation: www.baltimoreredevelopment.org
Bloomberg News: www.bloomberg.com
Boeing Company: www.boeing.com
Charter of the New Urbanism: www.cnu.com
City of Chicago: www.cityofchicago.org
Compuserve: www.compuserve.com
Council on Tall Buildings and Urban Habitat: www.ctbuh.org
Crime Reports: www.crimereport.com
Cvent Inc.: www.cvent.com
D.R. Horton, Inc.: www.DRHorton.com
eTurboNews: https://eturbonews.com
Family Watchdog Holdings, Inc.: www.familywatchdog.us
Fortune Magazine: www.fortunemagazine.com
High Line and Friends of the High Line: www.highline.org
Kentlands: www.kentlandsusa.com
Kriminologie-Lexikon ONLINE: www.krimilex.de
Microsoft: www.microsoft.com
National Building Museum: www.nbm.org
National September 11 Memorial & Museum: www.911memorial.org
NBC News (Ed.): www.everyblock.com
New Orleans Redevelopment Authority: www.growinghomenola.org
New York City Landmarks Preservation Commission: www.nyc.gov/landmarks
New York City: www.nyc.gov
Robert Fowler Retirement Media Inc.: www.55communityguide.com
Stanford University: www.stanford.edu
Starbucks: www.starbucks.com
U.S. Bureau of Labor Statistics: www.bls.gov
U.S. Census Bureau: www.census.gov
USA Business Review: www.businessreviewusa.com
Zillow Real Estate: www.zillow.com

Index

Printed in the United States
by Baker & Taylor Publisher Services